T0134430

DISCOVERING US

Sculptor Elisabeth Daynès recreates ancient human ancestors by applying forensic techniques to casts of fossil bones. Seen here, from left to right: *Australopithecus afarensis* from Hadar, Ethiopia, c. 3 million years old; *Homo floresiensis*, Liang Bua, Flores, Indonesia, 60,000–100,000; *Homo erectus pekinensis*, Zhoukoudian, China, 500,000–800,000; *Homo erectus sangiran* 17, Java, Indonesia, 1–1.3 million; *Homo sapiens*, Mandalson Mountain, Pyeongyang, North Korea, 10,000–12,000; *Homo neanderthalensis*, La Chapelle-Aux-Saints, France, 50–60,000.

DISCOVERING US

FIFTY GREAT DISCOVERIES IN HUMAN ORIGINS

EVAN HADINGHAM

FOREWORD BY KIRK JOHNSON

SIGNATURE BOOKS | 2021 | SALT LAKE CITY

www.signaturebooks.com

Design by Jason Francis.

FIRST EDITION | 2021

LIBRARY OF CONGRESS CATALOGING-IN-PUBLICATION DATA

Names: Hadingham, Evan, author. | Johnson, Kirk R., writer of foreword.

Title: Discovering us: fifty great discoveries in human origins / Evan Hadingham; foreword by Kirk Johnson.

Description: First edition. | Salt Lake City: Signature Books, 2021. | Includes bibliographical references and index. | Summary: "Over the past fifty years, researchers have made extraordinary discoveries that help us to understand who we are, where we came from, and what makes us human. Discovering Us brings our shared history to life and tells the stories behind fifty of the most important human origins discoveries ever made. Illustrated with stunning full-color photographs, this book celebrates science, exploration, and the search for what it means to be human. The Leakey Foundation is a non-profit organization formed in 1968 to fund human origins research and to share discoveries. Since then, the foundation has awarded more than 2,500 grants for research in 110 countries. Discovering Us highlights the thrilling fossil finds, groundbreaking primate behavior observations, and important scientific work of The Leakey Foundation researchers"—Provided by publisher.

Identifiers: LCCN 2021038625 | ISBN 9781560852773 (hardback)

Subjects: LCSH: Human beings—Origin. | Human evolution.

Classification: LCC GN281 .H334 2021 | DDC 599.93/8—dc23
LC record available at https://lccn.loc.gov/2021038625

CONTENTS

PREFACE

What is it to be "Human?" When the genome regions shared with us by chimpanzees are 99 percent identical to our DNA, what really makes us "Human?"

In a recent National Science Foundation poll, roughly half of all American adults said they believe that humans evolved from earlier forms of animals. That leaves roughly half of the American adults who don't. Many believe we have existed in our present form since the beginning of time. In a recent poll by the Pew Research Center, about 99 percent of scientists accept evolution as the way we became "Human." What causes the disconnect between the researchers and the non-scientists?

Often, the research that scientists conduct does not get passed on to the general public in a way that can be understood. Some people seem to think that the research is just the scientists' opinion rather than being based on careful observation and repeatable experiment. The general public often does not understand the methods used to support the scientific finds.

For more than 50 years The Leakey Foundation has been funding research, education, and public understanding of human origins, evolution, behavior, and survival. The Foundation funds research into early tool use and fossil studies, the geologic dating of those fossils, studies of ape behavior, modern hunter-gatherer groups, and how certain species have disappeared, or adapted to new environments.

Paleoanthropology is a relatively new science. When ancient fossil skulls were found in the 19th century looking similar to human skulls, they caught the imagination of scientists who wanted to know where they came from and how they related to humans. The science today is part geology, part archaeology, part biology, and a good part anthropology applied to what fossilized apes left behind, and what living hunter-gatherers are up to now. Behavior does not fossilize, hence the importance of studying our close primate relatives who may be doing some of the same things as our early ancestors.

An explosion in biological research and new biological information has come in the last 20 years. Much of the information has emerged from the study of whole genomes of different species, including *Homo sapiens.* Now, technology facilitates DNA and hormonal studies so researchers can learn how behavior and body chemicals are related.

In the 1960s, Dr. Louis S. B. Leakey often traveled to the United States to lecture and raise money for research projects. The Leakey Foundation was started by friends of Louis Leakey, many of them in Southern California, who were inspired by his ideas on human origins. These friends were lay scholars, homemakers, lawyers, bankers, film makers, and scientists in other fields. They knew that research would take years to complete, so they built an organization that would fund this science into the foreseeable future.

The group formed The Leakey Foundation in 1968 to support Louis and Mary Leakey's fieldwork and the groundbreaking research of young scientists, many of them, like Jane Goodall and Dian Fossey, embarking on unprecedented research that otherwise could not be funded. Since its early beginnings, The Leakey Foundation has been a leader in supporting women scientists.

In the 1970s, the early Leakey Foundation Board included scientists recruited from around the world by paleontologist Dr. F. Clark Howell, who recognized the value of bringing experts from many different fields to the study of human origins. *Row 1* (left to right): Barbara Newsom, Dr. Norton S. Ginsburg, Nancy Pelosi, Tita Caldwell. *Row 2* (left to right): Jane Goodall, Dr. F. Clark Howell, Kaye Jamison, Dr. Frederick Seitz, Ed N. Harrison, Joan Travis, Kay Woods, Dr. A. S. Msangi. *Row 3* (left to right): Dr. Irven Devore, Dr. Richard Flint, Coleman Morton, Dr. Richard S. Musangi, Dr. Melvin M. Payne, Edwin S. Munger, Gorden Getty, Dr Bogodar Winid, Mary Pechanec, Harold J. Coolidge, George D. Jagels, David A. Hamburg, Robert Beck, Lawrence Barker, Jr.

For example, Jane Goodall received 17 grants over the years for her work with the chimps in Gombe National Park, Tanzania.

A Science Executive Committee (SEC) of professors and field researchers was established by the Foundation to vet grant proposals. From the beginning, the SEC members have been among the best scientists in their fields. The Foundation often funds research for a Ph.D. or for a post-doctoral project. These early career grants make an important contribution to young scientists because they are awarded before the researchers have a well-known reputation and qualify for grants from the National Science Foundation (NSF) or other funders like National Geographic Society. This model has proven to be hugely successful, resulting in completed grants to over 2,000 individuals in 110 countries.

In an attempt to get the results of such cutting-edge science out to the general public, The Leakey Foundation has provided thousands of public lectures in natural science museums across the United States and has supported free school outreach lectures and workshops. In 2020, with the addition of our blog posts, web series, *Lunch Break Science*, and our award-winning podcast, *Origin Stories*, we have increased our reach tenfold over previous years.

As well as providing grants to young scientists, the Foundation has provided emergency funding to protect threatened primates

Fifty years after the founding of The Leakey Foundation, the Board includes trustees from across the United States as well as a Science Executive Committee of researchers who work around the world and are experts in anatomy, geology, genetics, primatology, paleontology, and behavior. *Row 1* (left to right): Dr. Diana McSherry, Chester Kamin, Don Dana, Nina L.Carroll, Dr. Kristen Hawkes, Camilla M. Smith, William P. Richards, J. Michael Gallagher, Dr. Alan J. Almquist. *Row 2* (left to right) Mark Jordan, Dr. Nina Jablonski, Jorge Leis, Julie M. LaNasa, Dr. Steven Kuhn, Dr. John Mitani, Dr. Martin N. Muller, Dr. Joan Silk, Jeanne Newman, Dr. Tom Plummer, Alice Corning, Janice Bell Kaye, Dr. Spencer Wells, Dr. Brenda Bradley, Owen O'Donnell, Dr. Carol Ward, Dr. Anne Stone, Dr. Daniel Lieberman, Dr. Craig Feibel, Cole Thomson.

and archaeological sites, seed money to start new sites, and bridge funding to keep long-term sites going until new funding can be found. Extended observations of primate societies across many generations have proven invaluable, particularly when combined with genetic data from the same groups.

For the 50th anniversary of the Foundation, we wanted to share 50 important discoveries in paleoanthropology in an easily understood way. We have highlighted significant milestones that have led scientists to increase knowledge of our species and of our ancient history. The story of our origins belongs to everyone. Research has told us so much about our early human stories; this is what Louis Leakey and the founders of The Leakey Foundation wished for as they sat at a kitchen table in Los Angeles and thought about how to keep this research funded. They wanted the Foundation to become a global institution that would welcome and support an ever-renewing group of scientists, reporters, and the general public just wanting to know what it is to be "Human."

Trustees of The Leakey Foundation are enthusiastic about the science, although they may not necessarily be scientists. Twenty to thirty individuals have volunteered to raise and invest funds and plan community outreach projects to keep funding the

The Foundation staff is small and efficient, organizing and funding grants, creating blogs, podcasts, science museum events, travel to research sites, and, during the pandemic, a series of online Zoom lectures on paleoanthropology. (Left to right) Arielle Johnson, Meredith Johnson, Executive Director Sharal Camisa, Grants Officer H. Gregory, Paddy Moore, Rachel Roberts.

science. They have traveled around the globe to see the research sites and meet the scientists.

The staff is similarly a small but enthusiastic group of people who love this science and spend creative time processing the grants and raising funds to keep the research going, as well as getting the news of discoveries out to our fellow humans.

In this book, our captivating author has told the stories of discovery in a most engaging way. Evan Hadingham is Senior Science Editor for NOVA, the science series produced at WGBH for the Public Broadcasting System (PBS). Over nearly three decades, he has helped oversee the production of award-winning NOVA documentaries, including many that drew on his interests in archaeology, prehistory, and human evolution—notably the series *In Search of Human Origins* with Don Johanson and *Becoming Human,* as well as single NOVA productions such as *Dawn of Humanity*, *Great Human Odyssey*, and *First Face of America*. His feature articles have appeared in *National Geographic*, *Smithsonian*, *The Atlantic,* and *Discover.*

This has been a collaborative project with The Leakey Foundation staff, many trustees, photographers, and scientists adding to the story. We are grateful to all who have made the story of *Discovering Us* so compelling.

Camilla Smith
President, The Leakey Foundation

FOREWORD

I can remember with great clarity when I first opened the pages of *National Geographic* and saw the image of Louis and Mary Leakey and their dogs in the excavation site at Olduvai Gorge. I was probably five years old and it was my first view of human paleontology in action. I returned to that picture many times as my interests in paleontology and archaeology blossomed into a profound obsession and, eventually, a long career.

This story speaks of the power of the *National Geographic* magazine and its century of imagery and story as one of the most accessible portals into the world of discovery. The continuously unfolding saga of the Leakey family and their collective passion to understand the origins of humanity in the Rift Valley of East Africa is one of the most compelling narratives of our time. And I suppose, by our time, I mean the time of the baby boomers and the post-war "great acceleration" of nearly everything. It has been said that 20 percent of all humans who have ever lived are presently alive. By the middle of this century we will approach a global population of 10 billion people and likely reach a point where our population levels out and begins to decline and deaths finally outpace births for the first time in human history.

Dr. Kirk Johnson, Sant Director of the National Museum of Natural History at the Smithsonian, visited Ellesmere Island in the Canadian Arctic in 2018. His experience in the Arctic was featured in *Polar Extremes*, a program in the NOVA series on PBS.

Most of the people presently alive will live to see this moment, making it not only unique in human experience but also in human awareness. In this same context, it becomes all the more important to understand the human story, from its evolutionary roots to its wholly unpredictable future.

As a paleontologist who found his first fossil at the age of 5, some 55 years ago, I have been reminded many times that the best fossils still lay in the ground awaiting discovery. Paleontology and archaeology, though they study ancient history and all of prehistory, are young disciplines with only a few hundred years under their collective belts. It is my experience that the way to make discoveries is simply to look for them, and I have seen endless examples of scientists who head to the field each season with valid optimism based on repeated success. In real terms, there are not that many field-based scientists and there is a tremendous amount of Planet Earth remaining to be searched. We know that new discoveries will be made and that is the root of genuine excitement.

I am charmed by the fact that our basic tools have not changed much in a century: field vehicles, shovels, pickaxes, hammers, awls, brushes, screens, wrapping paper, glue, and most importantly, a keen educated eye and endless patience. Of course, our modern century has added a myriad of other useful devices, concepts, and insights that allows us to accelerate our discoveries and increase our precision. From satellite imagery to global positioning devices and from mass spectrometer geochronology to the study and interpretation of ancient DNA, we can now interrogate old and new fossils in ways that were inconceivable 50 years ago.

In this light, it is possible to look back at the first 50 years of The Leakey Foundation and to realize that its founding was perfectly timed. In its first half a century, it provided critical support for a growing field of research through targeted grants. Now the science has matured, confronting its colonial origins and becoming more global.

As we watch the impacts of human-driven climate change ramify, we are reminded that our human ancestors lived and thrived through the tumultuous cool-cold cycles of the Pleistocene epoch over the last couple of million years. We have to peer deeper in time to the period when our primate ancestors thrived in the warm-hot cycles of the ice-free world of the Eocene epoch some 55 million years ago. What no mammalian human ancestor has ever experienced is the transition from an ice house world to a hot house world, since the last time that happened was more than 300 million years ago.

In this tidy volume, Evan Hadingham has boiled 50 years of The Leakey Foundation into 50 tales of discovery that range from key fossil discoveries to insights gleaned from the study of our nearest living relatives. For me, the last 50 years are clear in my memory and it is truly incredible to dip into this compendium and see how much we have learned. For me, it is a powerful dose of optimism about the power of science, discovery, and humanity.

Kirk Johnson
Sant Director, National Museum of Natural History

INTRODUCTION: THE LEAKEY LEGACY

On the morning of July 17, 1959, archaeologist Mary Leakey took her two Dalmatians for a walk in Tanzania's Olduvai Gorge. Her husband, Louis, who was recovering from flu, rested back in camp. Mary soon reached the site where he had first discovered prehistoric stone tools in 1931, a find that convinced him of Olduvai's potential to throw light on the riddle of human origins. Around 11 a.m., she glimpsed a bit of bone protruding from dirt on a nearby slope. Brushing aside some of the soil, she saw two big, black-brown teeth set in a massive upper jaw. The teeth immediately told her that she had found the skull of a human ancestor. According to the Leakeys' biographer, Mary rushed madly back to camp and cried out: "I've got him! I've got him! I've got him!" Groggy with fever, Louis asked, "Got what? Are you hurt?" "Him, the man! *Our* man," Mary said. "The one we've been looking for. Come quick. I've found his teeth!"[1]

Despite his illness, Louis leapt out of bed and they sped back to the site. Mary swept the rest of the soil away to reveal an almost complete, thick, robust skull. Louis's first reaction was disappointment. At first look, the enormous teeth actually resembled those of australopithecines, the extinct creatures that had been found in South African caves; although they were upright walkers, he regarded them as too primitive and ape-like to be part of the human family tree. A few hours later, continuing to study the huge skull back in camp, Louis spotted a few human-like details. He was seized with excitement. Now he was convinced that the new find, which he was to name *Zinjanthropus boisei*, was indeed a crucial part of the human story, "the

Louis, Mary, and 11-year-old Philip Leakey excavate a nearly 2-million-year-old activity floor at Olduvai Gorge in 1960, a year after their discovery of ***Zinjanthropus boisei*** opened a new chapter in the study of human origins in Africa.

Louis and Mary proudly display the reconstructed skull of *Zinjanthropus* (left), admired by President Julius Nyere of Tanzania (right). Time-Life's *March of Progress* illustration from 1965 (opposite) created a classic popular image of human origins.

connecting link between the South African near-men and true man as we know him."[2]

The discovery of Africa's "true man" ancestor became a worldwide sensation. A spellbinding communicator, Louis lectured about *Zinj* to packed halls in the U.S. and England, while *National Geographic*'s articles and films made the Leakeys a household name. *National Geographic* also gave them sorely needed funds to begin serious excavations at Olduvai. It was the start of the scientific saga of the Leakey family: a half-century of groundbreaking discoveries in East Africa, begun by Louis and Mary and continuing through their son Richard, Richard's wife Meave, and now Louis's granddaughter, Louise. Meanwhile, the impact of the Leakey dynasty has been multiplied many times over by The Leakey Foundation, which funds work at the forefront of fossil and primate studies and provides opportunities for up-and-coming scientists from developing nations. Originally established by friends of Louis Leakey in Southern California, The Leakey Foundation is no longer closely associated with the Leakey family but it maintains its founding mission, which was Louis's idea to fund early career research into human origins.

ANCESTORS ON PARADE

Why was Louis Leakey so diligently seeking the "true man?" And what are today's fossil hunters, the inheritors of the Leakey legacy, searching for? During the 1970s, the board of the recently established L.S.B. Leakey Foundation looked for an instantly recognizable image of human evolution that they could adopt for their logo. They chose one of the most famous scientific illustrations of all time, a drawing known as *The March of Progress*. Created in 1965 for the Time-Life book *Early Man*, it depicts a parade of 15 human ancestors striding purposefully from left to right, beginning with *Pliopithecus*, a small, knuckle-walking extinct ape from around 10 million years ago, and ending with a tall, proudly upright *Homo sapiens*. The artist, Rudolph F. Zallinger, carefully based his illustration on the best available fossil and archaeological evidence. While the Time-Life captions made it

clear that some of the ancestors in the parade were extinct, the image strongly implied that humanity's evolution followed a single, linear course, beginning with an early ape and ending with the triumph of modern humans. Zallinger was, in fact, skeptical of this oversimplified design, but the Time-Life editor insisted on it, rightly judging that it would have an enormous public impact.

Endlessly parodied in popular media—notably Homer Simpson's rise from *Monkius eatalotis* to *Homersapien*—the drawing annoys some biologists because they see natural selection, the mechanism that drives evolution, as essentially random. Acknowledging that the *March of Progress* had become "the canonical representation of evolution—the one picture immediately grasped and viscerally understood by all," Stephen Jay Gould condemned it as "false iconography" and added, "life is a copiously branching bush, continually pruned by the grim reaper of extinction, not a ladder of predictable progress."[3] Traits that help an animal survive may, over time, become more common and eventually lead to a new species branching off from an existing one. If conditions change, that trait or others may become liabilities and lead to a species dying out. No deliberate "march" is involved.

Nevertheless, when Louis and Mary Leakey were hunting for ancestral fossils at Olduvai Gorge in the 1950s and 1960s, the notion of our evolutionary path as an inevitable, single straight line projected far back into the past was not confined to the popular media. Based on the relatively scant fossil evidence then available, it was tempting to assume that there was something special about humanity's trajectory that, at a very early date, distinguished it from the rest of the natural world. This was the view of highly influential biologist Ernst Mayr, who forcefully advocated that human ancestry consisted of just three species, strung out in a single line of succession leading to *Homo sapiens* (much like *The March of Progress*). Because our invention of culture and stone tool technology had made early humans so successful, he argued, there was simply no room for other species to develop. If any competitors did emerge, Mayr believed, we deliberately wiped them out—including the Neanderthals, the population of robust humans that became extinct at the end of the last Ice Age. According to paleoanthropologist Ian Tattersall, Mayr's viewpoint was deeply appealing because it drew on a belief that "humanity has, like the hero of some ancient epic poem, struggled single-mindedly from primitiveness to its present peak of perfection."[4]

Nearly as recognizable as *The March of Progress* is the symbol of evolution as a branching tree. This visual metaphor for how one species relates to another was first sketched by Charles Darwin in a notebook from 1837, soon after he had returned from his historic round-the-world voyage in the *Beagle*. The famous diagram depicted his insight that any two related species can be traced back to a branching point, represented by their last common ancestor. The more closely related two species are, the more recent is their branching point and the closer together they are on the tree of life.

So which are humanity's closest relatives among other living creatures? In *The Origin of Species*, Darwin deliberately shied away from the question, figuring that it would only add fuel to the blaze of controversy ignited by his theory of natural selection. But in 1863, Darwin's staunchest supporter, naturalist Thomas Huxley, published evidence showing that modern humans are more closely related to African great apes than orangutans from Asia, and concluded that the ancestors of modern humans were therefore more likely to be found in Africa. In his 1871 book *The Descent of Man*, Darwin finally disclosed his views on humanity's origins and agreed with Huxley's argument that they were rooted in Africa. The advent of molecular genetics in the 1960s amply confirmed their conclusions. We now know that the genome regions shared by humans and chimpanzees are 99 percent identical; together with bonobos, chimpanzees are the closest branch to us among the tree of primates. Our branching-off point from their lineage is estimated to have been 5–7 million years ago.

Unfortunately, fossils from this period are particularly scarce, so we have only limited clues to what this last common ancestor was like. One thing we can be sure of is that it was *not* "the missing link" between humans and chimpanzees. As misleading as *The March of Progress*, "the missing link" is a persistent popular myth. Many people visualize it as a creature with a blend of modern human and chimpanzee features; in reality, chimpanzees as well as humans are constantly evolving. If you were to trace both lineages backwards in time, you would see each pass through many changes until they converged in their common ancestor.

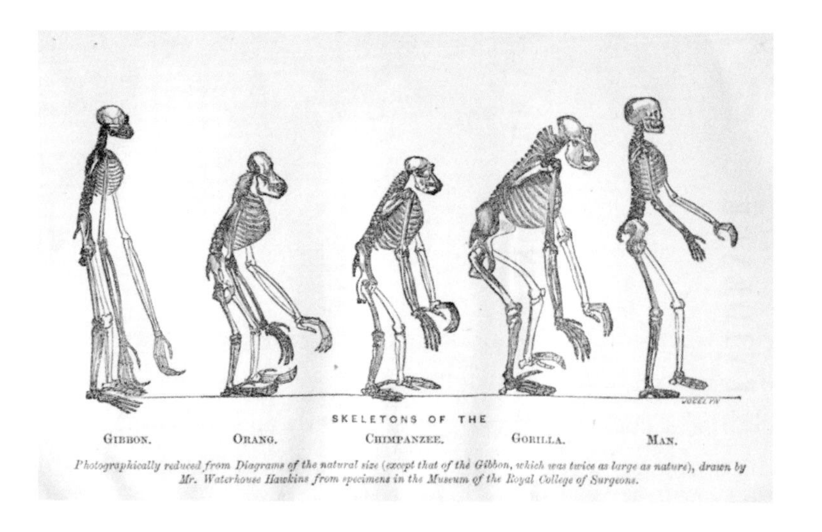

The march of humanity according to *The Simpsons* (top). An 1863 engraving from Thomas Huxley's *Evidence as to Man's Place in Nature* depicts the anatomical similarities between humans and the other great apes (bottom). Darwin's iconic "tree of life" sketch appeared in his 1837 notebook (opposite).

From this perspective, writes paleoanthropologist Don Johanson, each new find of a fossil human is "the discovery of another link in the long evolutionary chain from the common ancestor to modern humans today. They were all, each and every one of them, missing links."[5]

Spanning the 5–7 million years that divide us from the chimpanzee line are the many fossil ancestors described in this book. How should we visualize the tree that connects their story with ours? Did just one branch or several lead to us, and how many dead-end, extinct branches were there? Under Mayr's influence in the 1950s, the tree of humanity was dominated by a central trunk, with just a few dead side branches

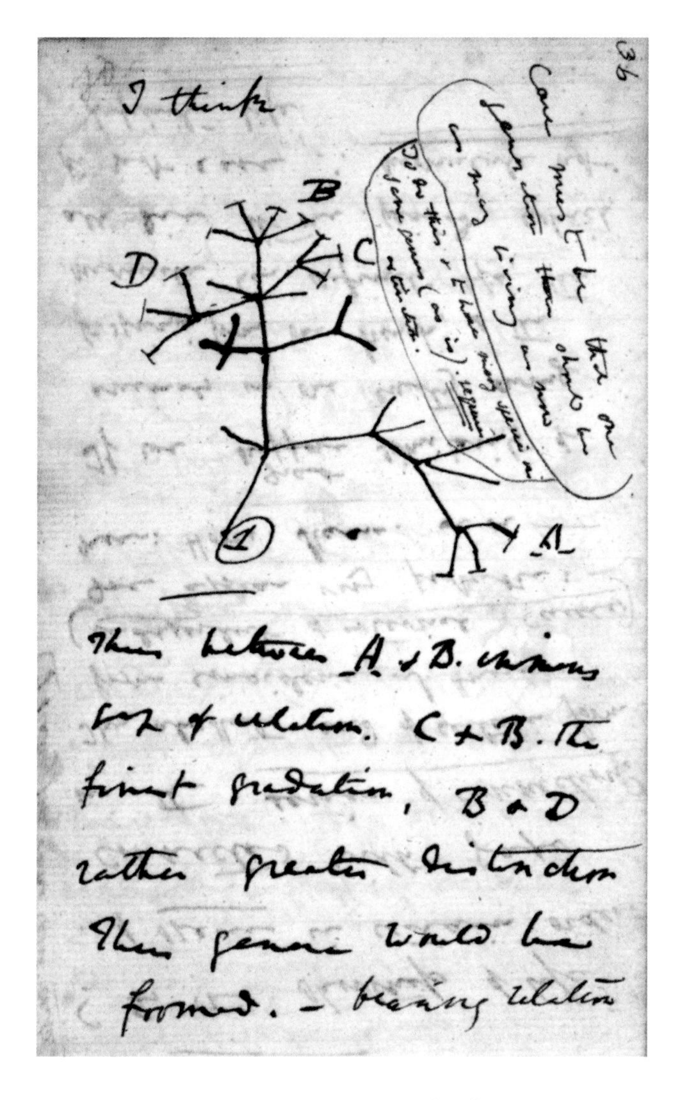

filled by extinct relatives like the Neanderthals. Even as little as two decades ago, the tree was relatively uncluttered; Ian Tattersall's chart from 1993 shows a dozen human species spanning 4 million years to the present. When he came to plot it again recently, it was far bushier, with double the number of species. Among the strangest of these recently added side branches are discoveries that sent shockwaves through the scientific community, like the tiny, meter-high *Homo floresiensis* or "Hobbit" of Liang Bua Cave on the Indonesian island of Flores (see p. 200), and the nearly 2,000 fossil bones of *Homo naledi* concealed in almost inaccessible crannies of South Africa's Rising Star Cave (p. 205). Both of these species have enigmatic combinations of small brains and modern-looking features, adding to the challenge of drawing connections among the thickets of dead ends on the tree.

In any case, the tree metaphor may have outrun its usefulness, thanks to the ever-growing complexity of human origins. Since 2010, a revolution in genome sequencing technology has unleashed an avalanche of new clues from ancient DNA. The technique involves drilling tiny samples from fossil bones in sterile conditions, similar to the "clean rooms" in which computer chips are manufactured, to prevent contamination from the scientists' own DNA. From these samples, entire human genomes can now be reconstructed. "High-throughput" machines are able to sequence millions of DNA fragments in parallel, dramatically lowering the time and costs of the analysis.

Since 2010, when the first five complete ancient human genomes were published, the technique has taken off. By 2020, the sequences of more than 5,000 individuals had been published. One of the first was a groundbreaking draft of the Neanderthal genome compiled by Swedish geneticist Svante Pääbo and his team, which overturned decades of consensus on the mysterious fate of these extinct European cousins from the Ice Age. Before Pääbo released his findings in 2010, researchers were confident that anatomically modern humans had replaced the Neanderthals—perhaps exterminating them—in the course of migrating from their African cradle and spreading across Europe around 40,000 years ago. Pääbo's teamwork showed that the modern DNA of most people outside Africa contains a small percentage—averaging two percent—of Neanderthal genes. This suggested repeated episodes of interbreeding between moderns and Neanderthals, most likely in the Middle East, as the main wave of our ancestors departed Africa. Scientists are now turning a fresh eye on the possibility that modern humans gradually assimilated at least part of their population, rather than wiping them all out.

Even more startling is another genome published in 2010, which testifies to the existence of a previously unknown major population known as the Denisovans, distinct from both

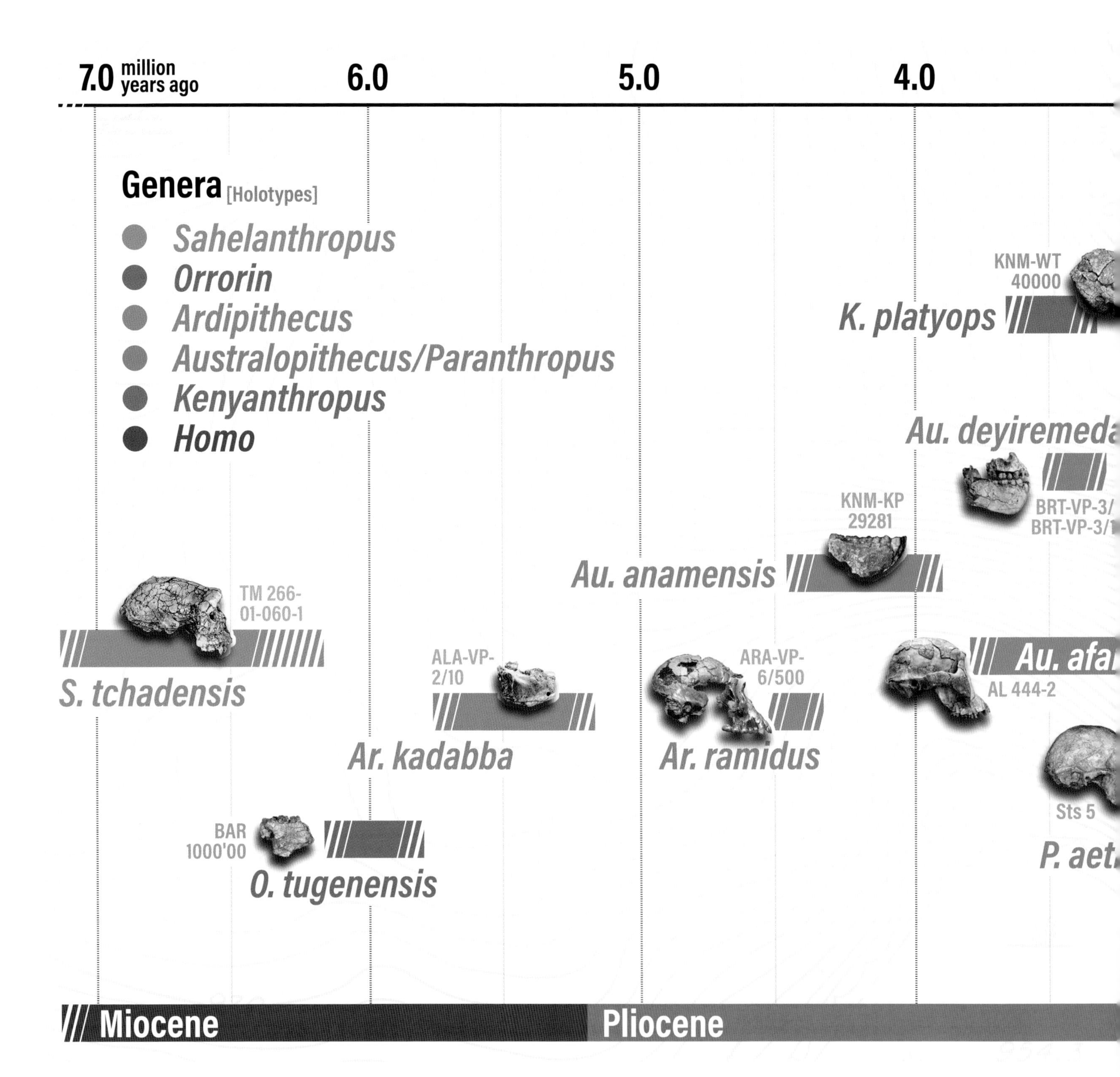

7.0 million years ago
6.0
5.0
4.0
Genera [Holotypes]
Sahelanthropus
Orrorin
Ardipithecus
Australopithecus/Paranthropus
Kenyanthropus
Homo
KNM-WT 40000
K. platyops
Au. deyiremeda
BRT-VP-3/
BRT-VP-3/1
KNM-KP 29281
Au. anamensis
TM 266-01-060-1
S. tchadensis
ALA-VP-2/10
Ar. kadabba
ARA-VP-6/500
Ar. ramidus
Au. afa
AL 444-2
Sts 5
P. aet
BAR 1000'00
O. tugenensis
Miocene
Pliocene

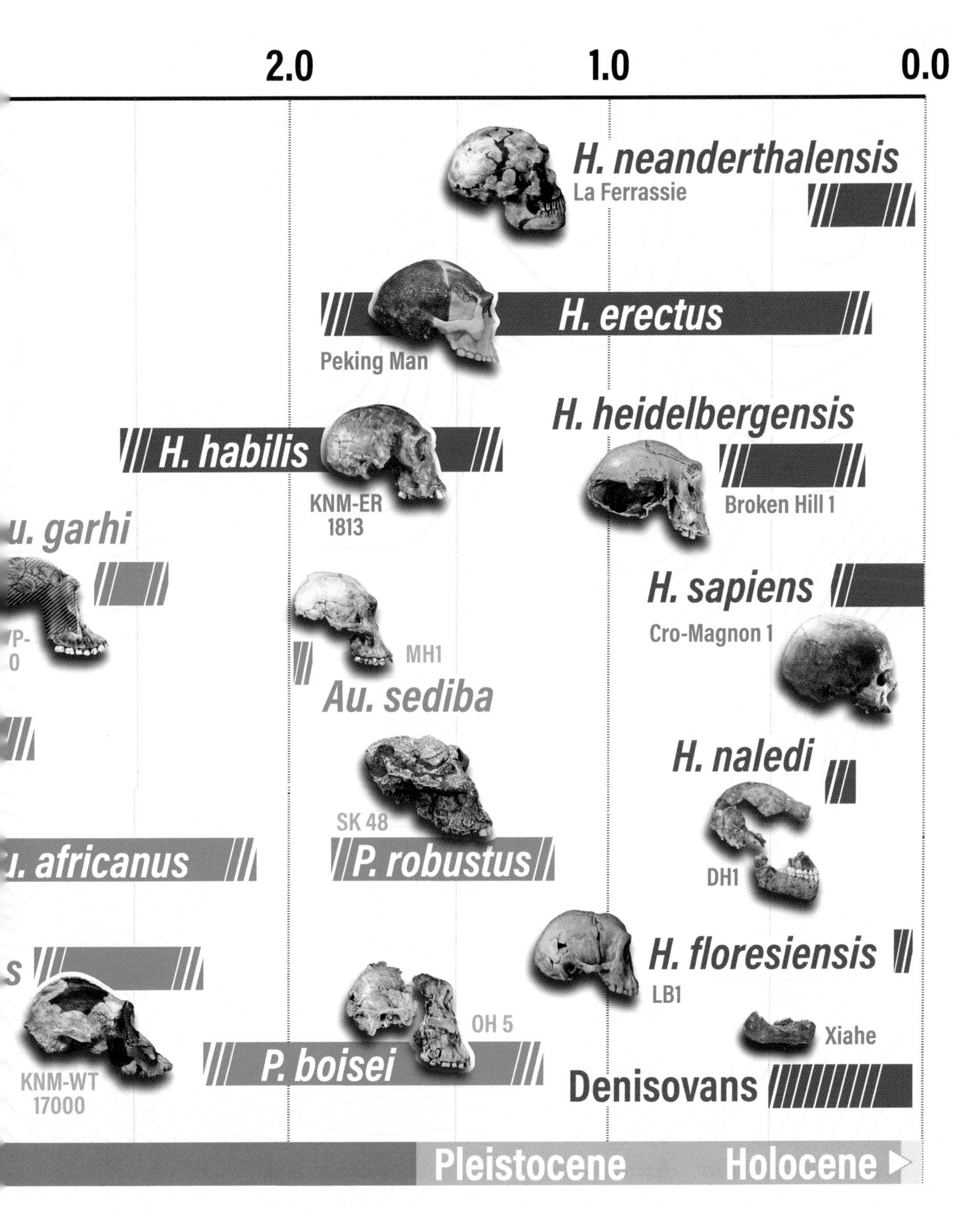

The emergence of humanity can no longer be visualized as a single-file march but as an intertwining bush or braided stream. This diagram considerably oversimplifies a picture that fresh discoveries are constantly redrawing.

Neanderthals and modern humans. Extraordinarily well-preserved DNA was retrieved from a single tiny pinkie bone, dated by soil layers that are at least 50,000 years old inside Denisova Cave in Siberia's Altai Mountains. The latest evidence now suggests that at least three different groups of Denisovans occupied huge areas of Eurasia and southeast Indonesia during the Ice Age, and that they interbred with both their Neanderthal and modern human contemporaries before dying out. Moreover, the Denisovan DNA indicates that they also mixed with at least one other "ghost population," another mystery group so far unidentified in either the fossil or archaeological record.

According to geneticist Turi King, these landmark genomic studies are uncovering "the tremendous degree to which populations globally are blended, repeatedly, over generations. Gone is the family tree spreading from Africa over the world, with each branch and twig representing a population that never touches others. What has been revealed is something much more complex and exciting: populations that split and re-form, change under selective pressures, move, exchange ideas, overthrow one another."[6]

One certainty emerges: while today we are the only human species on planet earth, our solitary condition is exceptional. For much of our evolutionary past, until about 40,000 years ago, our ancestors shared their world with several different hominin species. It is a very different picture of our evolution to Mayr's. Instead of a tree, some scientists now favor the image of a braided stream, with tributaries and interweaving rivulets, some petering out while others split off and then rejoin the main current. "Across the past 200,000 years," says anthropologist John Hawks, "these separate streams were swallowed up by the growth of one African branch of humanity. Humans spread through the world like a broad river delta, carrying slightly different fractions of the flow of ancient streams."[7]

With thousands of fossils added to the record each year and

(Opposite) Biochemist Matthias Meyer prepares an ancient DNA sample for analysis in a clean room at the Max Planck Institute for Evolutionary Anthropology, Leipzig. Completely sterile conditions are essential to prevent contamination by modern DNA. (Above) Louis Leakey points out the sites of his and Mary's greatest discoveries at Olduvai Gorge some three decades after their first field trip there in 1931.

the impact of ancient DNA only just beginning to be felt, the field of human origins is entering an exciting, dynamic phase. Many revelations lie ahead, since the main areas of human fossil finds so far explored in Africa, mainly in East Africa's Rift Valley and South Africa's limestone caves, make up a tiny fraction of the continent's terrain. In China and southern Asia, a new generation of scientists is opening up the vast potential of these equally underexplored regions. It is a time of ferment and excitement for young researchers, many supported by The Leakey Foundation, who have set out to make their mark in a fast-changing, challenging field. The continuing upheavals of the genomic revolution are a reminder of how little we know for sure, and the importance of questioning received notions. Sometimes, progress in science comes through maverick thinkers like Pååbo who are willing to look beyond the mainstream and push the field into fruitful new areas of discovery. Two other such independent minds belonged to Louis and Mary Leakey.

LOUIS LEAKEY: REBEL AND PIONEER

Louis's search for human ancestors in Africa had unpromising beginnings. When he led his first expeditions there in his twenties, the idea of an African cradle for humanity had fallen out of favor. A young Canadian physician, Davidson Black, had found an impressive skull of Peking Man in a cave near Beijing in 1929, which drew scholars' attention to a neglected earlier discovery of Java Man (both now thought to belong to the extinct ancestor known as *Homo erectus*). If you wanted to look for human fossils, most academics were convinced the place to go was Asia, not Africa. "There's nothing of significance to be found there," a Cambridge professor told Louis. "If you really want to spend your life studying early man, do it in Asia."[8]

But Louis Leakey was anything but a conformist. Born in 1903 near Nairobi, his parents were missionaries to the Kikuyu, Kenya's largest tribe. Louis grew up bilingual and was initiated as a tribal member, and he credited his Kikuyu identity for awakening his lifelong passion for observing the natural world. At age 13, he began picking up Stone Age tools on his hikes, igniting his curiosity about the past. Dispatched to an English public school at age 16, Louis was a restless misfit; at Cambridge University, he was branded as a showoff by straight-laced British students. In 1934, he shocked academic colleagues by leaving Henrietta "Frida" Avern, his pregnant wife, and Priscilla, his young daughter, to live with Mary Nicol, a talented archaeologist and illustrator. Two years later, they married, and Mary became a crucial intellectual partner and a meticulous archaeologist who made many important contributions of her own.

For the next two decades, Louis and Mary returned often to Olduvai, although with relatively little to show for their efforts until Mary spotted *Zinj* in 1959. While the discovery brought them fame and funding, Louis's enthusiasm for *Zinj* as a human ancestor didn't last long. Based on estimates from the skull, its small brain—at around 530 cc, not much bigger than a chimp's—made it an awkward fit for Louis's most fundamental beliefs about our origins, which were strongly influenced by Charles Darwin. In *The Descent of Man*, Darwin had suggested that the most obvious

Louis and Mary with geologist Peter Kent at Olduvai in 1935, a year before his marriage to her (left). Louis had a gift for communicating the enthusiasm of his quest for human origins to everyone he met (right). He experimented with ancient stone tools to skin a ram (opposite).

difference between us and the great apes, our ability to walk upright, was our ancestors' most crucial adaptation. Two-legged walking had freed the hands from locomotion, enabling them to become versatile instruments at the service of an expanding intellect; they could make weapons and throw them effectively. Darwin's arguments convinced Louis that the elusive ancestor he was searching for at Olduvai would be a big-brained toolmaker.

In 1960, less than a year since *Zinj*'s discovery, a much better candidate began to appear. Louis's teenage son Jonathan unearthed two skull parts and a lower jaw full of teeth, far more lightly built than *Zinj*'s rugged remains, a few hundred meters away. At over 600 cc, the new fossil's brain was significantly bigger. What's more, finger and foot bones were recovered that strikingly resembled our own. Here, surely, were the capable hands that had made Olduvai's stone chopping tools. Once center stage, *Zinj* was now demoted from the human family. Louis imagined he had been an intruding interloper at the toolmakers' sites, and had fallen victim either to the more advanced hominins or to another predator. In 1964, the new fossil was published under the apt name *Homo habilis*—or "handy man."

The academic establishment was at first staunchly skeptical of Louis's claim that he had found a new species of *Homo*, most of them believing that the australopithecines did, indeed, belong in the human chain linking us to the apes. A staggeringly early date for the *Zinj* skull poured fuel on the flames of skepticism. Louis had invited two Berkeley geologists, Garniss Curtis and Jack Evernden, to apply a novel dating technique to the ash from ancient volcanic eruptions that often interrupted the orderly sequence of soil layers in the Gorge. (The technique they developed involved measuring the slow decay of two radioactive isotopes in the ash.) They reported that the layer in which Mary had found the *Zinj* skull was 1.75 million years old—three times older than anyone, including Louis, had previously supposed.

The advent of scientific dating was a revolutionary advance that indicated a far longer timespan for humanity's evolution. Louis was now convinced that our *Homo* lineage must stretch back at least 20 million years into the past, with little change to our essential characteristics of upright walking and a big brain. At the time of his death in 1972, he still firmly clung to his belief that a single line of human ancestry connected *Homo*

(Left to right) Joan Travis, co-founder of The Leakey Foundation, with the so-called "Trimates" recruited by Louis Leakey to carry out the first long-term studies of wild primates: Biruté Mary Galdikas, Jane Goodall, and Dian Fossey.

habilis to *Homo sapiens*. He was also certain that other human relatives, including the australopithecines and *Homo erectus*, had flourished alongside *habilis* but had ultimately died out. "Man developed just like the animals did," he asserted, "with various species living side by side until the weaker died out or were annihilated, leaving the stronger until eventually man emerged."[9] Louis's vision of the diversity or "bushiness" of our family tree, which ran counter to what most other scientists then believed, was prophetic of our current understanding of human origins.

As well as his unorthodox views, Louis's flamboyant style made him an easy target for critics. While Mary patiently toiled on her meticulous excavations at Olduvai, resulting in major publications that have stood the test of time, Louis traveled the world, raising funds and enthusiastically promoting theories that were often based on intuition rather than solid evidence. Careless with his wardrobe, his personal eccentricities were hard for some to take, such as his alarming habit of driving continuously in first gear while reading the newspaper. His devotion to hobbies, ranging from birdwatching to code-breaking, bread baking, and raising tropical fish competitively, also raised eyebrows. "It annoyed a lot of people," paleoanthropologist Alan Walker noted, "who felt that with such a broad range of interests, he couldn't possibly be taking seriously their chosen field of study."[10]

But if Louis was sometimes distracted and creative with facts, his lifelong dedication to proving Darwin right had profound consequences. Mary and Louis's fieldwork at Olduvai nailed the case for humanity's African origins and opened up East Africa's Rift Valley as the staging ground for a half-century of spectacular fossil discoveries. Through his lectures, films, and personal encounters, his personal magnetism made an unforgettable impact on thousands. Archaeologist Rosemary Ritter said Louis "had a way of making even the littlest, most unimportant person, feel important. That's why people were so willing to work for him."[11] "Louis was a super-enthusiast," said Mary Smith, a *National Geographic* journalist. "He was so full of desire to know everything in the world that he just turned people on. He'd pour this electric energy into them as if they were vessels. Those who were receptive, he kept pushing and shoving and urging. And that's what sent the Goodalls to the jungles, the Fosseys to the mountaintops, and the Birutés to orangutanville."[12]

DISCOVERING THE GREAT APES: THE TRAILBLAZING TRIMATES

They were nicknamed the "trimates." Louis Leakey gave three young women—Jane Goodall, Dian Fossey, and Biruté Galdikas—separate missions to carry out the first long-term observations of great apes in their natural habitat. He was acting on a deep-seated hunch that the behavior of living primates could provide vital insights into the lives of our ancestors, and it was to prove one of his most enduring contributions. At a time when women were widely considered to be less capable scientists, all three faced daunting solitude, misogyny, uncooperative animals, and primitive camping conditions as they figured out the ground rules of an entirely new discipline of field primatology. Their landmark discoveries and passion for conservation continue to ignite new generations of young scientists today, inspiring them to dedicate their lives to observing and protecting endangered primates.

In addition, their work provides essential comparative evidence for interpreting the silent bones and stones of the fossil record. "No primatologist has ever claimed that humans were just like modern apes," notes Craig Stanford, a chimpanzee authority; instead, primate studies are "one obvious prism" through which to view our own behavior and that of vanished ancestors.[13] Discoveries such as the harmful physical effects of childhood deprivation among baboons (see p. 193) or the distinctive toolmaking styles passed down through chimpanzee groups (see p. 100) are not only fascinating in their own right, but also help us pose new questions and theories about the remote past.

The National Geographic Society was a major supporter of the Leakeys' and the "Trimates'" research. ***National Geographic*** magazine's vivid cover stories turned Jane Goodall (top), Dian Fossey (below left), and Biruté Mary Galdikas (below right) into world-famous celebrities.

(Opposite) Jane Goodall was indefatigable in her early efforts to track down and habituate the elusive chimps at Gombe. Louis Leakey gave her equally unflagging support and encouragement (left). Now in her eighties, Goodall is an internationally revered advocate for conservation and the environment (right).

JANE GOODALL AND THE ELUSIVE CHIMPS

In the spring of 1957, 23-year-old Jane Goodall, recently graduated from secretarial school in England, took a trip to Kenya to fulfill her childhood dream of "getting involved with animals."[14] In Nairobi, she contacted Louis Leakey to discuss her dream. Louis had been looking for someone to carry out a novel research mission: a systematic, long-term study of wild primates. Hiring Jane on the spot as his secretary, Louis eventually decided that she was the right person for his innovative project and began raising funds to support her fieldwork.

Finally, on July 14, 1960, Jane Goodall and her mother disembarked at the Gombe Stream Game Reserve on the eastern shores of Tanzania's Lake Tanganyika. Jane was venturing on a six-month effort to observe the behavior of wild chimpanzees in the wild, with no academic training or experience of living in a remote African forest. "I remember my first day," she recalled, "looking up from the shore to the forest, hearing the apes and the birds, and smelling the plants, and thinking this is very, very unreal. Then I started walking through the forest and, as soon as a chimp saw me, it would run away."[15]

Day after day, week after week, the pattern continued: she would hike up and down steep forested valleys and climb to the tops of hills, spotting distant chimps through her binoculars or catching fleeting glimpses as they fled from her approach. After more than three months of frustration and with the deadline for the end of her study looming, she had learned almost nothing about their behavior. Jane recalls writing to her sponsor, Louis Leakey, and saying, "'I can't do it.' Because he put all that money and trust in me and I was getting more and more worried. And every time I wrote back and said, 'I can't do it,' he'd write back, and his writing got bigger and bigger, saying 'I KNOW YOU CAN.'"[16]

Louis had sent the young Englishwoman to Tanzania in the belief that Gombe's lakeside setting would make a study of its wild chimp population particularly relevant for understanding ancient Olduvai. Besides the remains of *Zinj* and *Homo habilis* that Louis and Mary had unearthed there, they also found fossils of extinct forms of hippos, elephants, giraffes, and wildebeest, all indicating that 2 million years ago the arid gorge had been an abundant lakeside forest.

Another of Louis's hunches—that such a study would demand many months or years to yield results—was highly original. Before World War II, the idea of long-term observations of wild animals had barely gained a foothold. Most early expeditions consisted of collecting and describing dead primates rather than describing living ones. As a handful of field studies finally got off the ground in the early sixties, primatologists had little idea how to proceed. Sherwood Washburn, for example, says that "at that time, it was not a question of if you had a scientific report, but of whether you could see the animals at all. Before Jane, you were excited even if all you saw was an arm or backside disappearing into the bush."[17]

Jane's breakthrough at Gombe came while observing a chimp she had named David Greybeard after the distinctive white hair on his chin, who had taken to feeding near the camp. He was the first not to run away at Jane's presence. One day, she glimpsed David hunched over a termite nest, poking at it with a grass stem that he had stripped of its leaves. Then he extracted and licked the insect-laden stem. It was a rudimentary tool, and Jane knew right away what a momentous discovery it was; the most common definition of humanity, distinguishing us from the apes, was "man the toolmaker." In response to her telegram telling him the news, Louis sent back a memorable reply: "Now we must redefine man, redefine tool, or accept chimpanzees as humans."[18]

More groundbreaking observations followed as Gombe's chimps grew accustomed to Jane's presence. The textbooks said they were strictly vegetarian, but one morning Jane trained her binoculars on a chimp feasting on a wild piglet. Eventually, she recorded evidence that, singly or in groups, chimps hunted monkeys and other small animals on a fairly regular basis. And she began to accumulate insights into the bedrock of chimp social behavior, from the bonds between infants and mothers, to alliances between males and struggles for dominance or territory that sometimes spiraled into violence. Writing of Jane's achievement, primatologist Craig Stanford notes that "the idea that chimpanzees might be skilled toolmakers, cooperative hunters, or occasionally bloodthirsty killers, or that they might show the clear hallmarks of a

David Greybeard was one of Jane Goodall's favorite chimps at Gombe and was the first to habituate to her presence. In November 1960, Jane made her landmark observation of Greybeard using sticks to extract termites from their nest, the first recorded sighting of primates using tools. Certain communities of chimpanzees in West Africa's Taï Forest pass down traditions of using stone tools to crack open tree nuts, a favorite delicacy (opposite, see p. 100).

simple humanlike culture, sounded like science fiction in 1960. By the 1970s, it was textbook chimpanzee behavior."[19]

After more than a year of gathering unique field data, Jane was reluctantly persuaded by Louis that an academic degree would be essential for her work to be taken seriously. Exchanging the forests of Gombe for the cloisters of Cambridge University, she encountered stiff academic resistance. Giving chimps names like David Greybeard, Mike, Olly, and Fifi, and then attributing personalities and emotions to them was downright unscientific. "It was a bit shocking to be told I'd done everything wrong," Jane recalls.[20] "Everything. Fortunately, I thought back to my first teacher, when I was a child, who taught me that that wasn't true." That teacher was her dog Rusty, who had given her ample testimony of his personality. Jane's once heretical idea that studying primates requires empathy, rather than cool scientific detachment, is now taken for granted. Empathy, notes biologist Frans de Waal, "helps you to ask questions and to predict what your animals are going to do."[21]

Despite all the skepticism, Jane was awarded her doctorate in 1966 and she headed back to Gombe. Years of persistence and dedication started to pay off as her scientific papers and accumulating insights gradually established her reputation. By the late sixties, a new generation of graduate students came to Gombe to study under her, multiplying her impact. Jane's work set the standard for long-term chimp studies that now span Africa's equatorial forests, from Tanzania and Uganda in the Rift Valley all the way to Senegal in the far west. With each new field study, many supported by The Leakey Foundation, it becomes clearer that the behavior recorded at Gombe represents just one "flavor" in an extensive menu of chimp lifeways and cultures that varies from one region to another.

After publishing her definitive book *The Chimpanzees of Gombe* in 1986, Jane left active research and became a tireless public advocate for the protection of endangered primates. The outlook is not encouraging: between 2005–2013 alone, almost one-fifth of Africa's great ape population was lost due to

deforestation, the bushmeat trade, and diseases like Ebola. From more than 1 million chimpanzees that thronged the continent's equatorial forests a century ago, only around 170,000–300,000 are left. To combat the crisis, the Jane Goodall Institute works to promote environmental education, set up new sanctuaries, and strengthen efforts to stop illegal trafficking. In addition, her "Roots & Shoots" program develops imaginative projects with schools and youth organizations in nearly 100 countries to raise young people's awareness of conservation. "At the end of the day, I still think we can do it," Jane says. "Everywhere I go there are young people with shining eyes wanting to tell Dr. Jane what they are doing to make the world better. You have to be hopeful."[22]

It took Dian Fossey nearly three years to habituate the wild gorillas at the Karisoke Research Center that she founded in Rwanda. She trained and outfitted the National Park's team of rangers (above), more than 170 of whom have been killed in the last 20 years defending the gorillas.

DIAN FOSSEY, DEFENDER OF GORILLAS

The Leakey Foundation has provided vital support for long-term studies of mountain gorillas, which pose many daunting challenges. The majority live on the plunging, forested slopes of the Virunga Mountains, a spectacular range of mostly extinct volcanoes that marks the border between Rwanda, Uganda, and the Democratic Republic of the Congo. The beauty of the landscape belies the difficulty of tracking animals at 10,000 feet through forests persistently shrouded in clammy, cold mist.

For the second of the "trimates" recruited by Louis Leakey, Dian Fossey, the physical challenge of high altitude was compounded by her fear of heights and her emphysema, which was aggravated by a two-pack-a-day smoking habit. The Virunga gorillas proved particularly hard to habituate; it took three years for her to establish more than fleeting contacts with them.

Dian's task was also complicated by the inner demons of her personality. During the roughly 18 years that she spent in the mountains, she grew increasingly misanthropic. She resented outside helpers and tourists, whom she viewed as a threat to her work, and she often disparaged the local park rangers and officials in charge of protecting the gorillas.

But Dian's courage, determination, and scientific achievement have made her a legend, mythologized by the movie adapted from her autobiography, *Gorillas in the Mist.* Once again, Louis Leakey had recruited a woman with no academic training or fieldwork experience. According to *National Geographic* journalist Mary Smith who knew Louis well, he "trusted women for their patience, persistence, and perception—traits which he thought made them better students of primate behavior."[23] Dian Fossey had all those qualities. She studied a pre-veterinary course in college and directed an occupational therapy unit for disabled children at a Louisville hospital. In 1963, she took out a loan to travel to Africa, where she visited Olduvai Gorge and met Louis Leakey, and then traveled to Kabara, in the Virunga foothills. Here, in a remote meadow, was the place where one of Dian's heroes, naturalist George Schaller, had lived in a cabin and carried out a year-long study of mountain gorillas—the first ever attempted. Dian had a brief encounter with a gorilla group before she returned to Kentucky. Three years later, Leakey visited Louisville on a lecture tour, met Dian again, and was impressed by an article she had written for a local paper on her adventure in the Virungas. She quickly accepted his invitation to return to Kabara and dedicated years to unraveling the secrets of gorilla behavior and society.

Her mission got off to a disastrous start. After six difficult months at Kabara, a local rebellion broke out, and Zaire's military regime reacted with fierce reprisals against foreigners. Dian was marched down the mountain and held at park headquarters, where she was treated badly by her captors. After two weeks, she managed to fool them by finding a pretext to drive across the border to Rwanda. Undaunted, she set up a new camp on the Rwandan side of the Virungas and established the Karisoke Research Center.

From this base, Dian began her years of methodical observations of gorilla behavior, learning to recognize individuals by the distinctive patterns of their nostrils, or "nose-prints." She mapped their ranges, noted their calls and dietary habits, witnessed the killing of infants (usually by outsider males), and

Dian Fossey attracted criticism for the intimacy of her contacts with the Karisoke gorillas. But her fieldwork provided the first scientific accounts of wild gorilla behavior, including vital information on their social structure, reproduction, and life cycle. Her conservation of their skeletal remains at the Center (opposite) provided a foundation for research that still continues today (see p. 138).

traced subtle social patterns such as the migration of females between different groups. Persuaded by Louis to get her doctorate at Cambridge just as Jane had done, Dian authored a dissertation that, according to George Schaller, "really established the baseline" for understanding of the species.

On returning to Karisoke, her priorities gradually shifted away from collecting data to absorbing herself in the daily world of the animals she so closely identified with. According to author Virginia Morell, "Instead of observing objectively, Fossey, whose own human contacts were tenuous, plunged further into the life of the gorilla band. By the end, she had so thoroughly habituated the animals by imitating their behaviors that she was able to sit among them as if she were a gorilla."[24] This degree of intimate contact is shunned by primatologists today.

Dian also grew increasingly militant against poachers who killed gorillas and sold their stuffed heads for trophies and their hands for ashtrays. In a single four-month period, her African staff destroyed nearly 1,000 poachers' traps in the Karisoke research area. She targeted less malignant invaders, too, such as pastoralists forced by the unrest to drive their cattle up into the mountains, and hunter-gatherers who caught antelope in snares that would also trap and injure gorillas. Encouraging local people to think of her as a witch, she frightened them by wearing masks, casting hexes and spells, stealing and burning their possessions, and reportedly even whipping poachers with nettles. The killing and beheading of her most beloved gorilla, Digit, in 1977, was a turning point. "That tragedy practically unhinged her," says Mary Smith. "She became dangerous to herself and the Rwandans, because of her volcanic temper and her methods of interrogating alleged poachers."[25] In December 1985, she was brutally attacked and murdered in her cabin at Karisoke. The identity and motive of her killer are still unknown.

If Dian's tactics in the park were questionable, the Digit Fund that she established in memory of her favorite animal aroused

WELCOME TO
TANJUNG PUTING NATIONA
CAMP LEAKEY

A recent photo of Biruté Mary Galdikas taken at the entrance to Camp Leakey, the orangutan sanctuary in Borneo's Tanjung Puting reserve that she founded in 1971, when hardly anything was known about the elusive primate's biology and behavior.

worldwide concern about the mountain gorillas' plight. Renamed the Dian Fossey Gorilla Fund, it is now a leading effort among many initiatives that, together with enlightened management by Rwandan National Parks, has turned around the fortunes of Karisoke's gorillas. Rather than focusing only on enforcement, the Fossey Fund now supports community health projects and partners with local landowners and former hunters to protect endangered areas of the forest. The proof that cooperation rather than confrontation works can be seen in the steady rise in the numbers of mountain gorillas, which Dian feared would all be extinct by the year 2000. Instead, at the last census in 2018, their numbers totaled over 1,000, a 25 percent increase over 2010.

This success has come at a terrible human cost; over the past two decades, more than 170 park rangers sacrificed their lives to protect the gorillas. With violence in the DRC currently threatening to reach levels last seen in the civil war, Karisoke remains a fragile refuge that requires constant vigilance and outside support. At the moment, however, mountain gorillas remain Africa's only wild ape population known to be growing. This would not have happened without the unceasing efforts, if not the chosen tactics, of Dian Fossey, their most famous defender.

BIRUTÉ MARY GALDIKAS AND THE "IMPOSSIBLE" APE

The orangutans of Borneo and Sumatra are the most endangered of all the world's great apes, and perhaps the hardest of all to study. When the third Leakey "trimate," Biruté Mary Galdikas, told her UCLA professors about her mission, they thought it was impossible. "'You could be there for three years and not even meet an orangutan,'" she remembers one of them saying.[26] Mostly solitary, orangutans are tough to spot in the high forest canopy and accustoming them to the presence of humans is a lengthy business. "You habituate one orangutan at a time," Biruté explains. "And after you habituate one and he or she disappears, that's it. With chimpanzees or virtually any other primate, you habituate groups, so one individual may disappear but you still have the rest of the group. With orangutans it was one-on-one." Acclimatizing a single individual might demand a dozen years, she says.

Only a person of extraordinary resolve could have taken on such a challenge. But orangutans had fascinated Biruté since her childhood in Canada, and in 1969, while waiting in line to introduce herself to Louis Leakey at the end of a lecture at UCLA, she felt absolutely certain that he would help her to unlock their world. The 23-year-old Biruté had an academic background as a UCLA anthropology graduate, and was married to a photographer, Rod Brindamour, who was also a handyman, pilot, and experienced forester. These factors, combined with her determined attitude, quickly persuaded Louis that she was the right person to carry out the first long-term orangutan study, although it would take him almost two years to raise the necessary funds.

Biruté and Rod arrived at Borneo's remote Tanjung Puting reserve in 1971 and set up Camp Leakey, at first little more than a dank hut in the forest. There, they lived on rice and sardines, the constant rain reducing their clothes to rags. Wading through swamps to track the elusive apes, they faced run-ins with loggers, corrupt officials, and poachers who captured baby orangutans for the pet trade.

Despite all the obstacles, Biruté amassed thousands of hours of groundbreaking observations. "My first years in the field were years of discovery," she later recalled, "when merely finding a wild orangutan was exciting, when following an orangutan for a week or more was a triumph, when almost everything I learned about orangutans was new."[27] Among the revelations

(Above) Louis Leakey flanked by Biruté Mary Galdikas and her husband, Rod Brindamour, shortly after she persuaded Louis to send her to Borneo to carry out the first long-term study of wild orangutans—a formidable undertaking. She provided shelter for rescued and orphaned orangs at Camp Leakey (seen in 1965, opposite).

in her landmark 1978 thesis was the first evidence of the orangutan's unique reproductive cycle of 6–9 years, the longest birth interval of any mammal; during much of their long childhood, infants are inseparable from their mothers. Biruté's observations confirmed that male orangutans are indeed solitary animals, yet adolescent females often hang out together and males and females engage in lengthy courtships. She also documented the males' distinctive repertoire of calls and catalogued some 400 types of fruit, leaves, bark, and insects in their diet. With her academic reputation solidly established and her fame assured by a *National Geographic* cover story, Biruté successfully lobbied the Indonesian government to turn Tanjung Puting into a national park, and organized patrols to crack down on illegal poachers and loggers.

Eventually, all three "trimates" left the world of academic science to devote their lives to protecting and securing the future for endangered primates. In Biruté's case, she began her commitment to conservation within weeks of her arrival at Camp Leakey by confiscating a captive orphan named Sugito, which clung to her day and night. In 1979, Biruté and Rod divorced and he returned to Canada, partly so their three-year-old son, Binti, could have a more conventional upbringing. Since then, Biruté has increasingly focused on rescuing and caring for orphaned and captive orangutans. Today, over 300 rescued orangutans live in a rehabilitation center just outside the park, funded by the Orangutan Foundation International (OFI) that she started in 1986. In her autobiography, Biruté writes of her rescued apes as her forest family, "what we left behind to seek our destinies on the savannas and open lands of the earth."[28]

In the 1990s, investigative journalists painted a less romantic picture of conditions at Camp Leakey and the rehab center (neither of them supported by The Leakey Foundation). Biruté has stoutly denied these allegations. Meanwhile, wildlife experts have raised concerns about the practice of sending orphaned and ex-captive orangutans back into the wild. Over more than two decades, Biruté released at least 200 such animals. In 1995, the Indonesian authorities banned the reintroduction of ex-captives into the National Park, responding to worries that they might spread new diseases or increase competition for food and mates among wild populations already stressed by poaching and dwindling forest habitat. An additional concern is that interbreeding between different subspecies of orangutans might cause genetic abnormalities or fertility problems in their offspring.

The regulations do not apply to areas outside the Park, however, so OFI has worked with Indonesian officials to set up new release sites in other parts of the forest. Biruté's locally trained staff, now numbering around 200, patrols the sites and monitors the health of freed orangutans who return to visit feeding stations. Over 100 ex-captives have now been released at these new locations.

While experts debate the risks posed to wild populations by reintroduced animals, spiraling population losses present a bleak outlook for orangutan survival. A 2018 survey estimated that in the last 16 years, Borneo has lost some 150,000 orangutans—almost certainly more than the number now remaining. Their slow rate of reproduction compared to other primates puts them at special risk of extinction, while deforestation from the relentless spread of palm oil plantations destroys their habitat.

Almost single-handedly, Biruté succeeded in drawing worldwide attention to their plight, while her arduous field observations formed the baseline for a new generation of researchers, many supported by The Leakey Foundation, who now dedicate themselves to discovering the lifeways of this unique and intelligent ape before it's too late.

Together, Jane Goodall, Dian Fossey, and Biruté Mary Galdikas were at the forefront of primatology in the wild. They persisted in their missions in spite of extreme physical hardship and isolation, as well as open hostility from academic anthropologists who thought their focus on documenting animal lives was unscientific. As George Schaller commented, the "trimates" taught science that the great apes are "true individuals." "They have given us an empathy with our closest relatives, and that is the only thing that will save these animals in the end."[29]

For the Leakey family, the quest for human origins was a family affair. (Above) Mary Leakey shares her work with her sons Jonathan and Richard. At the Leakeys' camp at Koobi Fora, Kenya (opposite), Richard and Meave examine the skull of KNM-ER 1470 that became an international news sensation after its discovery in 1972.

MARY LEAKEY—"AN UNEQUALLED SAGA"

Richard Leakey took off from his field research station at Koobi Fora on the east side of Lake Turkana, Kenya, on September 26, 1972. He flew south to Nairobi with a precious cargo carefully packed in a wooden box. From Nairobi, he drove 12 miles to the family house at Langata, walked into Louis Leakey's office, opened the box, and placed an astonishing fossil skull into his father's hands.

For the last three weeks, Meave Leakey, a paleontologist married to Richard, had been painstakingly piecing together an intricate puzzle of 150 tiny bone fragments that Bernard Ngeneo, one of the expert team of Kenyan fossil hunters known as the

Hominid Gang, had collected from the slope of an eroded gully.[30] She had not quite finished assembling the skull when Richard insisted on taking it with him to Nairobi. "And I remember saying, 'But Richard, I can't bear to be parted,'" recalls Meave, "'because here it's almost together and I can't get it'—'No,' he said, 'I have to take it because I have to show my father.'"[31] In fragile health, Louis was leaving the next day for a lecture tour. Instinctively, Richard knew this might be Louis's last chance to see the new find.

The skull seemed like final proof of what his father had passionately argued for decades: that a large-brained human ancestor had strode across the grasslands of East Africa more than 2 million years ago side by side with other, now extinct, hominin species. With its flat, broad face and high-vaulted skull, it appeared to be the best candidate yet for an ancient forerunner of *Homo.* "The feeling of delight and enthusiasm [on Louis's face] as he handled this skull was almost tangible," his assistant Pat Barrett remembered, "and his hands, and the expression on his face can never be forgotten."[32] The announcement of the skull, known as KNM-ER 1470, would propel Richard to worldwide fame, including the cover of *Time* magazine and a BBC television series. As they celebrated that evening at Langata, an intimate, rare moment of amity reigned between father and son after months of tension. Five days later, Richard got the news that Louis had died of a heart attack in London.

Louis Leakey excavates at Olduvai (left) and Mary Leakey at Laetoli, Tanzania (right). Their persistence and dedication, together with contrasting temperaments and skills, resulted in a brilliant partnership.

After Louis's passing, his life and achievements were widely commemorated in the popular and scholarly press, but his family's contributions to science were far from over. A few days after his funeral and memorial service in Nairobi, Mary Leakey returned to Olduvai Gorge to renew her investigations of the sites that had made them household names. If Louis was the flamboyant visionary and showman, Mary was the patient, painstaking scientist—a contrast that archaeologist J. Desmond Clark noted. "Louis often joked of 'Leakey's Luck,'" Clark wrote, "and indeed this could have been a factor, but the most significant part of the 'luck' was Mary. Louis was full of energy, charisma, and innovative planning, but it was Mary who

produced the data as a result of her meticulous methods of excavation and recording work in the field."[33]

The year before Louis died, she had published a landmark report on the excavation of the earliest layers at the Olduvai sites that had yielded the famous fossils of *Zinj* and *Homo habilis* as well as evidence of stone toolmaking. These layers spanned an era from about 1.8 to 1.2 million years ago. She interpreted those containing scatters of bone and stone as "living floors," marking human activity. Exposing them required monumental effort; the *Zinj* site, for example, involved removing hundreds of tons of cliffside soil and rocks with picks and shovels. Then, once Mary's team encountered the cement-like layers containing the living floors, they switched to dental picks, soft paintbrushes, and fine sieving of the soil so they would not miss even tiny fragments of stone and bone.

Ultimately, Mary catalogued 3,150 large fossil bones and thousands more fragments, together with 2,470 stone tools and around the same number of stone flakes and chips, all scattered across successive living floors of an area only a little larger than a tennis court. Previous generations of fossil hunters would have merely dug holes through deposits to retrieve impressive specimens, acquiring little or no information about their context. By contrast, Mary's team exposed broad horizontal surfaces and recorded the exact position of finds, so her data preserved potentially vital associations between fossil bones and stone tools. This held out the possibility of reconstructing the behavior and activities of ancestors like *Zinj* at the site, although Mary's cautious, critical outlook made her wary of jumping to hasty conclusions and interpretations, and she often scorned the way other scientists indulged in colorful theories and elaborate claims based on slender data. But the precision of her recording techniques opened the door for others to apply new ideas and tests to the wealth of evidence from Olduvai.

For example, since Mary's team had saved tiny scraps that would normally have been thrown out, later researchers found

(Opposite) Mary Leakey begins uncovering the skull of *Zinjanthropus boisei* at Olduvai in 1959. In an interview, Meave Leakey recalls that Mary "was meticulous about how she did everything and it was her careful looking that found *Zinj*...Her excavations were so carefully recorded, so carefully done, inch by inch...and those methods are basically the same as what we use today."

they could refit bone fragments together and show how animal long bones had been cracked open with the help of stone tools to extract the protein-rich marrow inside. Researchers also studied cutmarks on the bones left by sharp stone flakes as the Olduvai hominins stripped them of meat. Sometimes these cutmarks appeared on top of previous gnawing marks left by other carnivores, fueling prolonged debates about whether purposeful hunting or furtive scavenging had been more important for our ancestors' survival. Regardless of that debate, Mary's practice of precisely mapping the associations between stones and bones marked a fundamental advance. According to archaeologist Rick Potts, it was "as exciting now as it was then, because it means that the hominids were engaged in cooperative behavior of some sort. They also had sites they returned to on a regular basis, although we don't know what the magnet was."[34]

Following Louis's death, Mary devoted herself to analyzing the more recent, upper layers at Olduvai, which dated from around 600,000 to just over 1 million years ago. Looking back at the totality of layers at the site, she documented only slight improvements in stone technology, notably the introduction of hand axes, which were probably all-purpose tools that required more advanced crafting of the stone. But for more than one million years, the Olduvai hominids had repeated more or less the same basic forms for the stone tools on which they relied for survival.

Mary's analysis was an extraordinary achievement. Gathered together in five monumental volumes, her Olduvai reports set a new benchmark in archaeological science and are still referred to by specialists today who investigate the diet, activities, and environment of our early ancestors.

With the analysis of the Olduvai excavations largely behind her, it was time for Mary to move on. During her first trip to Africa with Louis back in 1935, a local informant had tipped them off to a location called Laetoli, 30 miles south of Olduvai, where he recalled seeing "bones like stone." Briefly exploring the site, they found a number of fossils sandwiched by deposits of volcanic ash. Now, four decades later and in her early sixties, Mary assembled a team to investigate Laetoli's full potential.

At the start of her second season in 1976, the team found a promising variety of hominin teeth and skull fragments. However, nothing prepared them for the unique discovery that followed: thousands of animal trackways on a surface dating back over a million and a half years before the earliest levels at Olduvai. Some prints looked as fresh as if they had been made yesterday, preserved by a layer of soft, fine volcanic ash that a rain shower had hardened into rock soon after the animals passed by. These animal tracks were remarkable enough, representing more than 20 species, from rhinoceroses and giraffes to hares and insects. Then, just a few days before they were due to pack up camp, four unusual tracks were spotted—this time, resembling the footprints of upright, bipedal walkers. By the end of the 1978 season, two parallel trails of footprints, about 70 in all, had been unearthed, probably left by three australopithecines walking at a leisurely pace across the landscape. (For details of the footprints, see p. 82.) In her autobiography, Mary Leakey recalled that, at first, the discovery seemed "something so extraordinary that I could scarcely take it in or comprehend its implications for some while."[35]

At around 3.6 million years old, the Laetoli footprints are the world's earliest. They still provoke astonishment, capturing a few fleeting moments in the lives of our distant ancestors, and were stunning proof that upright walking had been one of our species' oldest adaptations.

They were also the crowning achievement of Mary's four-decade dedication to applying innovative and painstaking science to uncovering the human past. Eventually, facing health issues and

The next Leakey generation: Richard and Meave Leakey with daughters Louise and Samira at their Lake Turkana home in 1977.

tiring of the solitude and difficult field conditions in Tanzania, Mary retired to the Leakey family home at Langata in 1983. From then until her death in 1996, she completed the Olduvai publications, wrote her memoirs, and found time to visit friends and her ten grandchildren. Hers was a lifetime of accomplishment no less significant than Louis's. "Their personalities were different," J. Desmond Clark remembered, "Louis exuberant and outgoing, Mary quiet with infinite patience and attention to detail: the combination and mutual compatibility was their success. Their life and work together is an unequalled saga in the field of prehistory."[36]

THE QUEST CONTINUES: RICHARD AND MEAVE

At first, Richard Leakey struggled to establish his own path in the face of his parents' fame. One of his earliest memories is of complaining about the heat, thirst, and boredom he felt at age six as they toiled away beside him at a fossil site. Exasperated, his father told him to go find his own bone. Secretly pleased to have something to do, he spotted one and started brushing away the soil. What emerged was a sizeable jaw, complete with shiny teeth. "I'd found what turned out to be the complete jaw of an extinct pig," Richard remembers. "I was thrilled by the discovery and thrilled even more so by the excitement it evinced in my parents."[37] But then they took over the find, leaving him "furious and deeply upset."[38] Richard credits his anger at the incident with the stirrings of an early determination to have nothing to do with his parents' world of archaeological digs and fossil hunting.

In his teens, Richard dropped out of high school, learned how to fly a plane, and started a business taking tourists on wildlife photo safaris. Eventually, he found work organizing fossil

Richard Leakey examines a fossil with colleagues from the "Hominid Gang" at Koobi Fora. From left to right: Kamoya Kimeu, Richard Leakey, Bernard Ngeneo, Wambua Mangao, and Harrison Mutua.

hunting expeditions for the National Museum of Kenya, and his attitude to his parents' field slowly changed. At age 23, he led the Kenyan contingent of an international expedition organized by his father to explore new fossil territory along Ethiopia's remote Omo River. Here his team found parts of two skulls dating to around 200,000 years ago, among the earliest examples of fully modern humans in Africa. It was the first of many groundbreaking discoveries.

But on the Omo trip, he resented the way that Louis, back in Nairobi, still viewed him as a safari organizer rather than an increasingly knowledgeable scientist. The opportunity to strike out on his own soon came. On a flight back to Omo from Nairobi, a storm diverted his plane around the little explored eastern shore of Kenya's Lake Turkana, the world's largest permanent desert lake. From the plane, Richard spotted exposures of sedimentary rock that looked promising for fossils. Borrowing a helicopter from the U.S. team at Omo, he touched down briefly in the area and saw stone tools and animal teeth strewn everywhere he stopped to look. By 1968, he had established a base camp at Koobi Fora and recruited the legendary, highly skilled Hominid Gang as his team. "I was excited to have finally found a site with fossils where neither of my parents had ever been before," Richard says. "It wasn't their show; it was my show. That gave me a lot of enthusiasm."[39]

Turkana's abundance exceeded all expectations; "one of our biggest problems was the sheer quantity of fossils!" Richard wrote.[40] Among the most momentous finds was the 1984 discovery of Turkana Boy, a 1.6 million-year-old skeleton of a *Homo erectus* youth. The Hominid Gang's Kamoya Kimeu was the first to spot a single matchbook-sized piece of skull lying

camouflaged against volcanic rubble. When Richard and anatomist Alan Walker began investigating, they found that the roots of a thorny acacia bush had grown straight through the ancient face. The rest of the skeleton had tumbled down a slope, requiring another five years of sleuthing and excavating to recover it. With 80 percent of its skeleton still intact, even delicate rib bones, Turkana Boy remains one of the most complete ancestral skeletons ever found and a crucial focus of debate about how our modern human anatomy took shape (see p. 122).

In 1989, after nearly two decades of these and many other stunning discoveries at Turkana, Kenyan president Daniel Arap Moi invited Richard Leakey to head the nation's Wildlife Service. It was a courageous career switch, requiring him to tackle an epidemic of elephant and rhino poaching. His tough crackdown on the illegal ivory trade made him many enemies, and when he lost both legs in a plane crash in 1993, sabotage was widely suspected. Further government service followed, and he rose to Cabinet Secretary under President Moi. When elephant poaching worsened again, Richard returned to head the Wildlife Service again at age seventy.

Throughout Richard's years of public service, Meave Leakey continued to direct the fieldwork at Koobi Fora and add to the phenomenal record of fossil finds there. One of her team's most important discoveries was along a dry riverbed at a place called Kanapoi one day in May 1994, when the blueish glint of fragments of tooth enamel embedded in what looked like a rock caught the eye of Hominid Gang member Wambua Mangao. Alerted by his shouts, Meave hastened to the spot, turned over the rock, and realized that it was the fossilized remains of the upper jaw of an extinct ape. From its location sandwiched between volcanic ash layers, she knew that the jaw must date to around 4 million years old. A few days later, Kamoya Kimeu

Meave and Louise Leakey (opposite) have continued their family's extraordinary record of fossil discovery at the Turkana Basin Institute, founded by Richard in 2005 to involve African scholars and the local community in human origins research and education.

spotted more bones, including the two ends of a shin bone or tibia. The shape of the tibia's base strongly indicated that the fossil ape was a straight-legged, upright walker. Overall, the Kanapoi finds added up to a distinctive blend of features not seen in the famous Lucy skeleton discovered in the Ethiopian region of the Rift Valley by Donald Johanson in 1974 (see p. 79). The evidence was enough for Meave to announce the team's find as a new species, *Australopithecus anamensis,* after the name for "lake" in the local Turkana language. It vindicated the conviction of both Louis and Richard that multiple types of hominins had flourished side-by-side like twigs on a bush, rather than the lingering view of a single, deep-rooted tree leading to humanity. Striking confirmation of this view came in 2016, when a team in Ethiopia led by Leakey grantee Yohannes Haile-Selassie unearthed a remarkably complete fossil face of *A. anamensis,* looking very different to Lucy's more ape-like appearance.

Meanwhile, a new chapter of the Leakey saga began when Meave and Richard's daughter, Louise, joined their efforts. In 1993, she had just started college in England and was still uncertain if she wanted to dedicate her life to arduous fieldwork. Then came news of her father's plane crash, and she was abruptly called back to Kenya to run the Turkana field camp single-handedly in her parents' absence. That experience helped change her mind. Eventually receiving her doctorate at University College, London, she became a vital co-leader and partner in Meave's continuing discoveries in the Turkana Basin. Today, Louise's own research focuses on the Basin's ecology and on another episode of striking diversity among fossil hominins around 2 million years ago, the time that our own genus, *Homo*, was finally emerging. And like her grandfather, she reaches out to a wide public with her talks and interviews on human origins and conservation issues.

THE FUTURE OF THE PAST: EXPANDING THE LEAKEY MISSION

In the late 1960s, Louis lectured frequently on college campuses, connecting the human past with the human future, and expressing his concerns about the global environment. But he charged a pittance, for example, asking Caltech for $100 for one such event. Eventually, a group of Louis's friends and supporters in California grew increasingly worried by his failing health and the constant stress of raising funds for his field projects and the "trimates." Hoping to bring in more money and relieve Louis from mundane logistical chores, they began organizing a foundation, quickly discovering that running one without an endowment was a challenging proposition. At one crisis point, just $3.00 was left in the Foundation's spending account.

After many ups and downs, the L.S.B. Leakey Fund for Research Relevant to Man's Origins opened its doors on March 26, 1968. Its stated aims were to explore and excavate sites related to human evolution, support lab analysis of the resulting finds, fund studies of living primates, and draw insights into contemporary human behavior. The Leakey Foundation also aimed to support the careers of young African scientists through fellowships, scholarships and travel grants, and to communicate the fruits of his research activities to the public through lectures, symposia, and films. A $1 million matching grant by Robert Beck, a wealthy retired electronics engineer and businessman, finally set the Foundation on an even keel, and it has thrived ever since.

More than five decades later, the Foundation remains true to its original vision—increasing scientific knowledge, educational and public understanding of human origins, behavior, and survival. Its wide range of projects include such leading areas of scientific inquiry as the human microbiome, longevity and child development, the genomes of ancient and living people, and the biology and behavior of some of the world's most endangered primates.

(Opposite) Accurate dating of the Leakeys' fossil discoveries in East Africa was crucial to their scientific impact. The late geologist Frank Brown of the University of Utah (at right) developed a chemical technique of analyzing distinct ash layers laid down by ancient volcanic eruptions, which enabled fossils to be pinpointed in time. His Utah colleague, geologist Patrick Gathogo (at left), uses similar analyses of Rift Valley sediments to reconstruct shifts in climate and volcanism over the last 4 million years.

During its half-century history, The Leakey Foundation helped underwrite the early careers of distinguished experts such as Zeray Alemseged, Donald Johanson, and Yohannes Haile-Selassie. Today, it offers support for promising scholars and students in nations with a wealth of ancestral heritage but limited educational opportunities. The Franklin Moser Baldwin Memorial Fellowship gives such students a chance to secure further education or training from an institution abroad, helping them take on leadership roles in the research they carry out in their home countries. A few of the latest generation of Baldwin Fellows are profiled here (see p. 40).

Meanwhile, a separate new initiative is supporting young scholars from Kenya, Tanzania, and Eritrea who want to pursue projects in the earth sciences related to the search for human origins. The Francis H. Brown African Scholars Fund was named after Professor Brown, a geologist who dedicated more than half a century to establishing a timeline of the rock and ash layers in the Rift Valley; this provided a vital way to estimate the age of the ancestral fossils discovered there. Closely involved with the Leakeys' work at Olduvai Gorge and with the Foundation's mission, Brown started the Fund from a conviction that Africans should have the same opportunities as Europeans and Americans to investigate their shared human ancestry.

A similar motivation led to The Leakey Foundation's third educational initiative, the Joan Cogswell Donner Field School Scholarship. This enables students in developing countries to gain a foothold in the essential fieldwork techniques required to investigate fossil sites.

As well as nurturing the careers of budding paleoanthropologists, The Leakey Foundation plays a critical role in enabling the study and protection of vulnerable primates, many on the brink of extinction. Since 1960, the Foundation has given more than 2,500 grants to primatology sites in Africa and the Americas. One of them, the Lomas Barbudal Monkey Project in Costa Rica, continues to reveal astonishing cultural behavior among capuchin monkeys more than a quarter of a century after its director, Susan Perry, received her first grant to begin observing them (see p. 228). Perry says that "most, if not all, of today's long-term primatology projects owe their origins and/or existence to the support of The Leakey Foundation."[41] Since primates are long-lived, extended observations are often essential to a full understanding of their behavior and development.

Bonobo social life, for example, follows a strikingly unusual social pattern unlike that of any other great ape, yet their scarcity, remote rainforest location, and the turmoil of the war-torn Democratic Republic of the Congo have all hindered scientific investigation. One of the few such initiatives is the LuiKotale Bonobo Project, which has been carrying out continuous studies of two habituated neighboring bonobo communities, one begun in 2004 and the other in 2011. The Project's director, Barbara Fruth, credits The Leakey Foundation with making such long-term monitoring possible. "Of particular interest during that funding period," Fruth says,[42] "were the births of six and the death of two infants, as well as of one adult. These were important observations contributing to our understanding of the life history of bonobos. Moreover, the two habituated communities experienced dramatic shifts in their relationship, including patterns of range use." These shifts might have escaped notice in a shorter-term study, Fruth believes. The LuiKotale and other projects reveal an intriguing picture of bonobo society, dominated by powerful females who have extended relationships with female friends outside their own group. Their shifting communities and generally peaceable lives stand in stark contrast to the violent,

territorial, male-centered realm of chimpanzees (see p. 107). Multi-generational studies like these have paid off not only in landmark scientific findings but in raising public awareness, for example, of the gorillas at the Karisoke Research Center in the Virunga Mountains founded by Dian Fossey, and the chimpanzees at research sites such as Gombe and Mahale in Tanzania, the Taï Forest in Côte d'Ivoire, and Kibale in Uganda.

Finally, The Leakey Foundation recognizes the vital importance of engaging the public in the process of science; indeed, it believes that scientific work is incomplete unless it is widely shared. It continues to develop innovative and imaginative ways to reach broad audiences, from its National Speaker Series lecture program to the award-winning *Origin Stories* podcast. As Leakey Foundation grantees help shape our ever-changing understanding of human origins, its outreach spreads an exciting and inclusive vision of the place of humans and our primate cousins in the natural world.

The modern science of human origins that the Leakey family and the "trimates" set in motion has done much to debunk the centuries-old image of humans as the pinnacle of creation, somehow separate from the natural order. The work of a half-century of Louis's successors has shown how intimately the human story is bound up with the forests and savannas from which our ancestors emerged millions of years ago. As we catch glimpses of ourselves mirrored in the behavior of living primates and in the fossil record of vanished ages, the lessons we draw about extinction and resilience may yet prove essential for our survival.

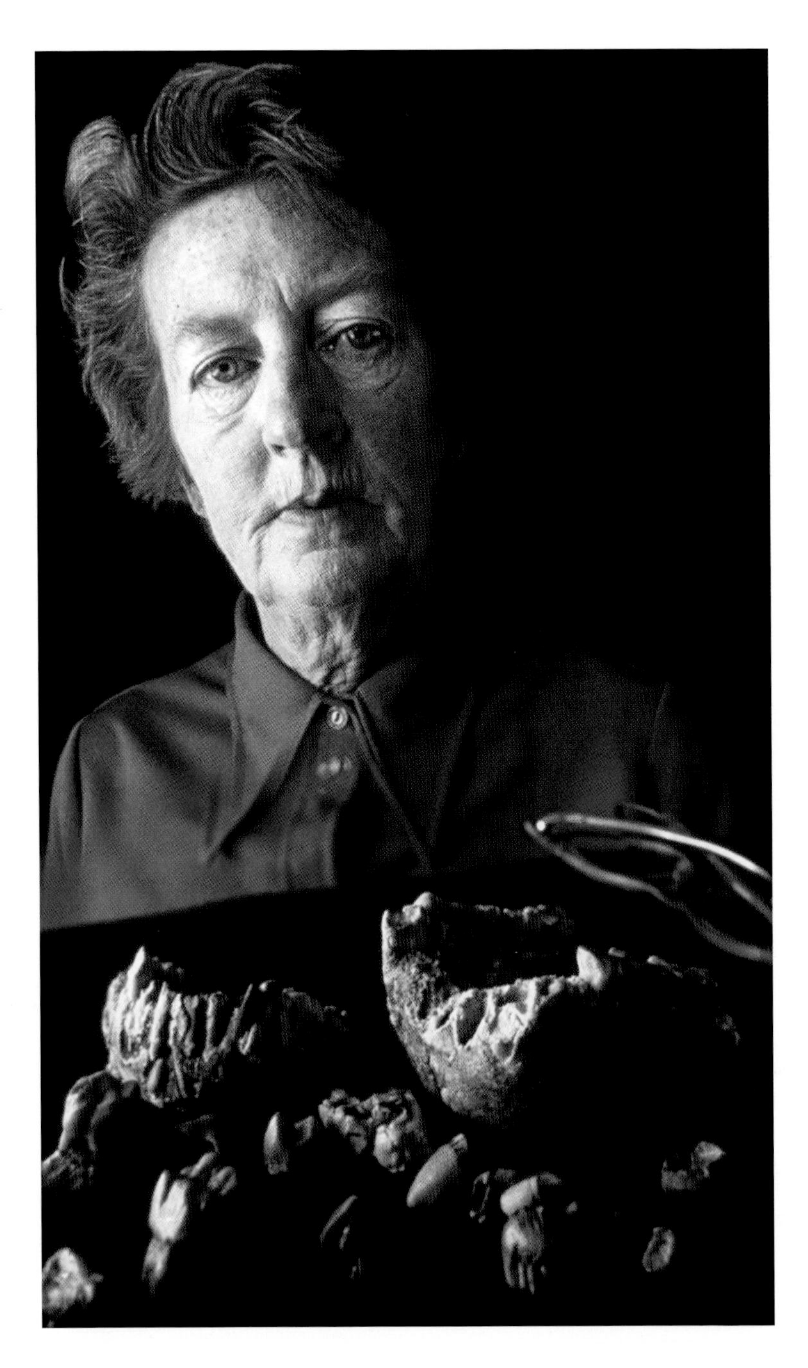

(Above) Mary Leakey at Laetoli with hominin fossils that she discovered at the site. (Opposite) Louis and Mary were "the pathfinders," wrote J. Desmond Clark, "whose achievements are an inspiration and a challenge to those who follow on today and in the continuing work on increasing the clarity and precision of the record of our human past."

SUPPORTING A NEW GENERATION OF SCIENTISTS

Since 1978, The Leakey Foundation's Franklin Mosher Baldwin Memorial Fellowship has helped scholars and students, many of them from developing nations, to obtain education or training outside their home countries. Often, these nations have a wealth of prehistoric heritage but lack educational opportunities for students to pursue human origins research. By enabling bright young scholars to obtain graduate education, the Fellowship helps equip these individuals to assume a leadership role in the future of paleoanthropology and primatology. Recent Baldwin Fellows have come from as far afield as Algeria, Ethiopia, Tanzania, South Africa, India, Iran, Indonesia, and China.

NICO ALAMSYAH

A Baldwin Fellowship enabled Nico Alamsyah (at right) to study for an advanced degree in archaeological science at Lakehead University in Canada. His research focuses on analyzing thousands of animal and human bones excavated at Liang Bua, the cave on the island of Flores, Indonesia, celebrated for the extraordinary discovery of *Homo floresiensis*, a previously unknown branch of humanity better known as the Hobbit (see p. 200). Nico plans to return to his position at Indonesia's National Research Center for Archaeology, where he hopes to set up a lab dedicated to reconstructing ancient diets and environments.

PENINA EMMANUEL KADALIDA

A Baldwin Fellowship was "a dream come true" for Penina Emmanuel Kadalida, seen on the right taking a "selfie" with a statue of Louis Leakey outside the Kenya Museum. The Fellowship enabled her to study for a paleoanthropology doctorate at the University of Minnesota. Her research will focus on investigating the evidence for early paleolithic toolmakers at a newly discovered site in northern Tanzania. After graduating, Penina plans to return to the University of Dar es Salaam with the ultimate hope of establishing an institute that will conduct fieldwork and provide training for Tanzanian students to engage in groundbreaking projects that have traditionally been carried out by outsiders.

SHARMI SEN

A primatologist investigating one of her field's most intriguing controversies, Sharmi Sen works in Ethiopia's rugged Simien Mountains, where she is contributing to a long-running study of gelada monkeys. As in a number of other highly competitive primate societies, gelada males sometimes kill the infants of unrelated females. Why some males practice infanticide and others do not has long been debated (see p. 232). Currently studying for her doctorate at the University of Michigan supported by a Baldwin Fellowship, Sharmi is using DNA analysis and other techniques to reveal clues such as the paternity of gelada infants. She hopes eventually to start her own long-term field project in her home nation of India.

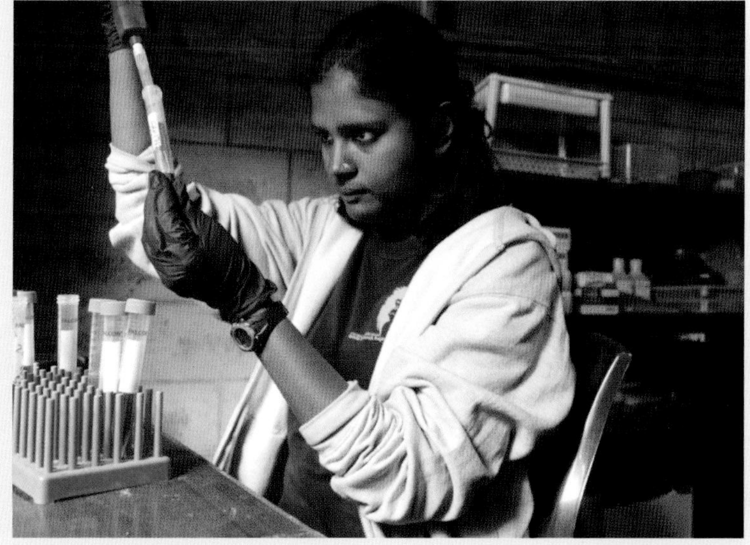

ONJA RAZAFINDRATSIMA

Currently on the Biology faculty at UC Berkeley and a National Geographic Explorer, Onja Razafindratsima is an ecologist with a passion for tropical ecology and conservation. For over a decade, partly supported by a Baldwin Fellowship, she has studied Madagascar's iconic lemurs and their fruit diet, which is crucial for the dispersal of seed plants across the island's endangered rainforests (see p. 57). Onja feels strongly about involving local communities and Malagasy students in her work. "Seeing a Malagasy scientist such as myself leading a research project in Madagascar," she says, "young Malagasy students might be inspired to pursue a biological field."

HESHAM SALLAM

Although less celebrated than its pyramids and mummies, Egypt has some of Africa's most important paleontological sites, including fossil beds preserving early ancestral primates from the Eocene era. Until recently, these sites have only been dug sporadically by outsiders. In 2010, Hesham Sallam founded the Mansoura University Vertebrate Paleontology Center (MUVP), the first paleontology lab in the Middle East, where he has trained a large team of postgraduate Egyptian students to carry out fieldwork. A Baldwin Fellowship supported the completion of Hesham's doctorate at the University of Oxford. "As the founder of MUVP," he says, "I can confidently say that the Baldwin Fellowship has played a critical role in helping me to build the first professional vertebrate paleontology program in Egypt."

ROSEMARY ANN BLERSCH

Partly supported by a Baldwin Fellowship, Rosemary Ann Blersch recently completed her doctorate at Canada's University of Lethbridge. Her research focuses on the health and behavior of vervet monkeys in South Africa's semi-arid Karoo desert, particularly the question of how infectious disease and parasites spread among their populations. Given rapid climate change and large-scale human migration, understanding how infectious disease spreads among wildlife and its potential link to human epidemics is becoming increasingly vital. Blersch hopes to establish biological anthropology as an academic discipline and to provide opportunities for South African students to work in her field.

DISCOVERING US

PIONEERING PRIMATES

A stunningly intact skeleton, unearthed by a farmer in China's central Hubei province in 2003, provides a vivid snapshot of one of countless species of early primates that thrived before the emergence of the human family. Weighing only one ounce and fitting easily in the palm of a hand, *Archicebus achilles* was hardly an imposing creature. Even more diminutive than today's smallest primate, the pygmy mouse lemur, it had a little head and eyes; a five-inch tail that was longer than its body; grasping hands and feet; and long hind legs suitable for acrobatic leaps between branches. Chris Beard, a paleontologist who helped analyze the fossil, says that *Archicebus* "looks like an odd hybrid with the feet of a small monkey, the arms, legs and teeth of a very primitive primate."[1] The discovery of this tiny beast and its unexpected mosaic of features is the strongest evidence yet that the earliest known anthropoids—a branch of the primate tree that would give rise to today's monkeys, apes, and, ultimately, us—originally emerged in Asia.

In the mid-1800s, Charles Darwin and Thomas Huxley observed anatomical resemblances between chimpanzees, gorillas, and humans, and argued that we must all share a common ancestor. Darwin was also the first to suggest that humanity's roots would be found in Africa, and the search for primate origins was long ruled by a similar assumption. A rich array of fossils, collected over three decades by paleontologist Elwyn Simons in Egypt's Fayum desert, seemed to plant the idea on a solid foundation. Then, in 1992, Beard discovered some tiny fossil jaws and teeth in China's Jiangsu Province that dated from around 45 million years ago. Naming the species *Eosimias*, or "dawn monkey," Beard argued that it pointed to Asia, not Africa, as the birthplace of primates. While Beard's case was soon strengthened by other finds from China and Myanmar, the humid tropical climate of that epoch, the Eocene (about 56–34 million years ago) preserved fossils so poorly that most discoveries consist only of partial scraps of bone and teeth. *Archicebus*, on the other hand, was half-complete. Unfortunately, the tiny fossil was also squashed flat, obscuring many vital details of its anatomy.

Specialists at the European Synchrotron Radiation Facility in Grenoble, France, took high-resolution scans of the skeleton. An international team, led by paleontologist Xijun Ni, reconstructed it virtually in three dimensions and spent five years assessing its place in the anthropoid lineage. The 55 million-year-old *Archicebus* precedes Beard's 1992 discovery in the adjoining province by 10 million years. Its appearance coincides with the start of the Eocene and an exceptional burst of global warming, when crocodiles swam in Canada and palm trees grew in southeast Alaska. Dense tropical forests, lush with the fruiting plants on which primates thrive, spread into northern latitudes. As these early primates flourished and diversified, they acquired some of their characteristic adaptations to life in the trees: an enlarged brain compared to other mammals, enhanced stereo vision, nails instead of claws, and grasping hands and feet. Beard imagines *Archicebus* as "a kind of frenetic animal, anxious and agile, climbing and leaping around."[2]

But if the first primates did emerge in Asia, how did they spread across the rest of the globe? During the Eocene, both Africa and South America were island continents, separated from Asia by vast ocean expanses. Yet by 38 million years ago, species similar to Asia's *Eosimias* had appeared in Africa, and monkey-like close relatives showed up in South America not long afterwards. Improbable as it might seem, a widely accepted theory is that ancestral primates migrated from Asia by taking an accidental ocean voyage on floating trees and mats of vegetation, a process that naturalists aptly term "sweepstakes dispersal." While such events are indeed as rare as hitting the jackpot,

The exquisite fossil of one of the earliest known primates, *Archicebus achilles*, was found by splitting apart a rock with remnants of the skeleton (left) and its mirror-image imprint (right). This minuscule one-ounce creature was a forerunner of today's monkeys, great apes, and humans.

Tarsiers (opposite) are forest-dwelling nocturnal primates in Indonesia and Borneo. *Archicebus* had tarsier-like features but the tiny eyes of a daytime hunter. The vast diversity of ancient primate ancestors across 55 million years is represented in a mural at the American Museum of Natural History (above).

travelers have witnessed violent storms tearing up chunks of land and vegetation from riverbanks, which then drift out to sea. There are even reports of upright trees still standing on natural rafts, an important detail supporting the theory that primates could survive the trip. According to one simulation, a drift voyage across the mid-Eocene Atlantic (then less than 1,000 miles wide) could take two months, which is probably too long for even the hardiest monkey. But with the right wind in the "sails" of a standing tree, it might last as little as two weeks.

While the theory of "sailor monkeys" may strain belief, the evidence for it has steadily grown, including fossils of distinctive rodents and birds that were fellow passengers with the primates. It is astonishing to think that the riotously diverse New World monkeys—capuchins, tamarins, marmosets, spider, and howler monkeys—may all owe their existence to a handful of seagoing castaways. No less amazing is the bigger primate saga, at times equally contingent on the lucky survival of small populations. This odyssey spans across 50 million years, encompassing the miniature *Archicebus* in its Asian homeland to Lucy in the African Rift Valley and countless other species, and, eventually, us.

ALESI: BABY APE

The long road to the discovery of Alesi, an exquisitely preserved 13-million-year-old fossil of a baby ape, begins with a 17-year-old student sitting in a museum auditorium in Kenya, listening to Richard Leakey. Isaiah Nengo recalls how Leakey's talk "mesmerized" him.[3] Six years later, he wrote to the Leakeys, who invited him to volunteer at the National Museums of Kenya. Nengo's assignment was to help organize the Museum's collection of fossil ape remains from the Miocene Epoch (roughly 23–5 million years ago). Although the great apes are today restricted to just a few species and regions in Africa and Asia, a profusion of species flourished across vast areas of the Old World during this 18-million-year epoch, which saw the emergence of the common ancestor of all living great apes and humans. Frustratingly, this wealth of Miocene ape species is mostly represented by mere scraps of jaws and limb bones, and as Nengo sorted through these fragments in the Museum, he became hooked on the challenge they posed. Figuring out the relationships between different branches of the ape family tree is essential to understanding how our earliest ancestors eventually arose in Africa.

Nengo earned his doctorate at Harvard before returning to Kenya, determined to take on the Miocene challenge in the field. "All the links between us and living apes were to be found in the Miocene," he says.[4] "Where did that common ancestor emerge in Africa? And what did it look like?" In 2014, with The Leakey Foundation support, Nengo organized a two-week survey of a promising area known as Napudet, west of Lake Turkana. Lushly forested in the Miocene, Napudet today is an empty, treeless wilderness. As the end of the two weeks loomed, his team had found nothing at all among the barren outcrops. Trudging despondently back to camp on September 4th, Nengo's colleague John Ekusi spotted what he thought was the knee bone of an elephant. Yet the minute he brushed off the fossil, Ekusi could see it was the top of a tiny primate skull. In seconds, despondency turned to jubilation and, as they meticulously uncovered it over the next couple of days, to astonishment at its miraculously intact condition.

Described by paleoanthropologist Ellen Miller as "the size of a lemon," and on the outside remarkably similar to that of a baby gibbon, Alesi is one of the most complete skulls of an extinct ape ever found. "It's also just shocking to find a skull from a baby," Miller adds, "because the bones are so delicate, the chance of having something like that fossilized is astronomically small."[5] And Alesi turned out to be just as extraordinary on the inside. When Nengo took the skull to the state of the art x-ray facility in France—where *Archicebus achilles* was also scanned—the high-powered, three-dimensional x-ray images showed internal structures that are almost never preserved. Hidden inside the

(Opposite) The nearly complete skull of a 13-million-year-old ape, *Nyanzapithecus alesi,* soon after its discovery in northern Kenya in 2014. It represents rare fossil evidence from the Miocene, when the common ancestor of today's great apes and humans emerged.

jaw, Alesi's unerupted adult teeth are visible, complete with minute enamel growth rings laid down each day like tree rings. The total of the rings adds up to 485 days, or just over 16 months: the measure of baby Alesi's fleeting lifespan.

Even more amazing, the images reveal Alesi's incredibly delicate bony inner ear tubes, which would have helped her balance as she made her way through the Miocene forest. They are close to those of Africa's living great apes, which move more cautiously through the trees compared to gibbons with their wild acrobatic swings. This special mixture of great ape and gibbon features gives Alesi her distinct identity, and brings us a step closer to unearthing one of many possible missing roots at the base of our family tree.

Isaiah Nengo, now Associate Director of Kenya's Turkana Basin Institute, finds deeper lessons in his study of Alesi. "The story of our becoming human," he writes, "is of ape species thriving in Africa—a petri dish for evolutionary experiments on ape-human form, conjuring up variety after variety over millions of years. If anything, a better understanding of our ape origins and our relationship to our few surviving cousin species could inspire us to reflect on our true place in nature."[6]

(Left) 3D x-ray images of Alesi's tiny skull are scanned at the European Synchrotron Radiation Facility in Grenoble, France. The images revealed such intricate hidden details as the brain cavity, inner ears, and unerupted adult teeth. (Opposite) Kenyan fossil hunters Akai Ekes and John Ekusi watch as paleoanthropologist Isaiah Nengo lifts Alesi's fossil skull from the sandstone bedrock after six hours of excavation.

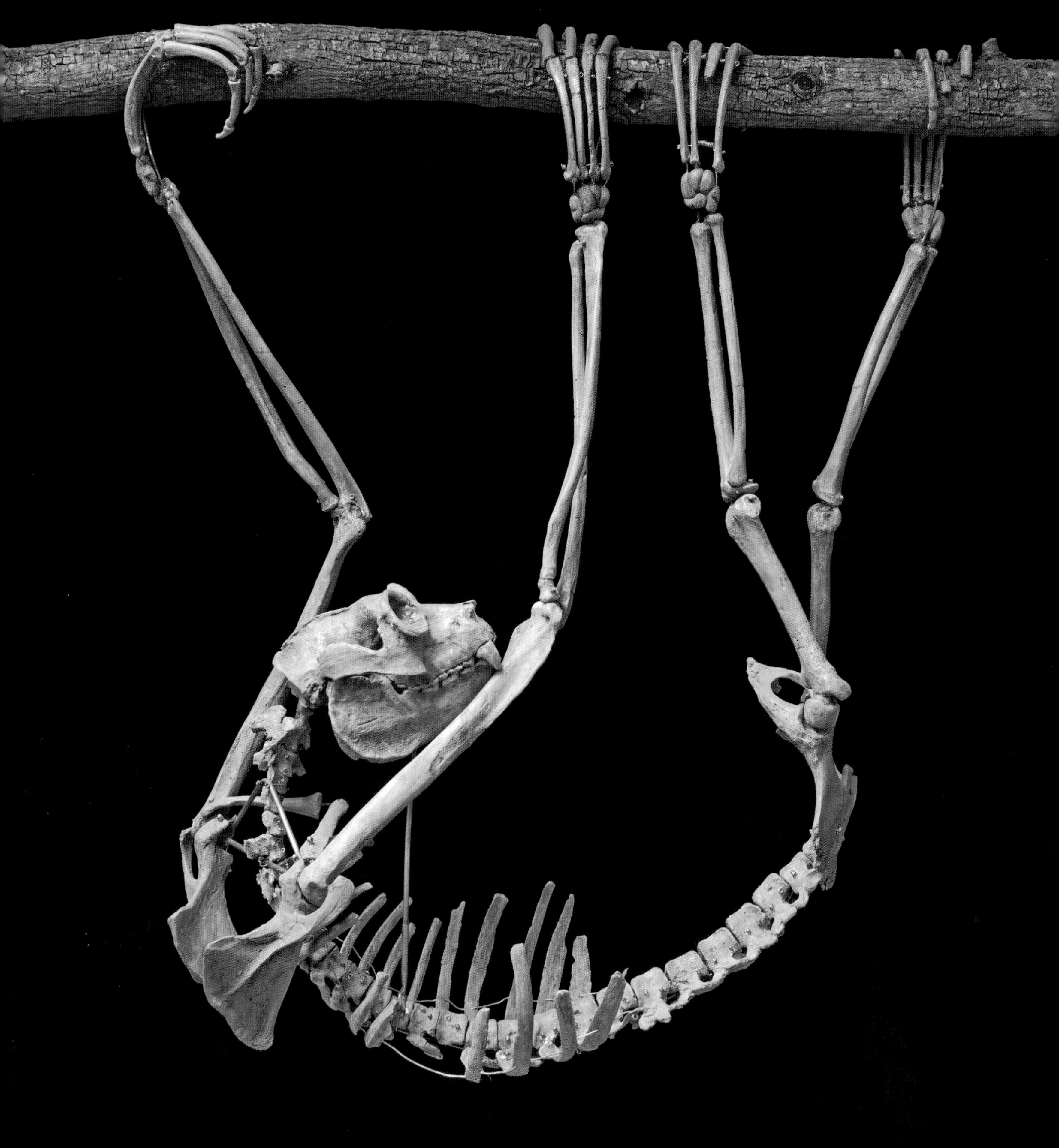

MADAGASCAR'S LOST GIANTS

A single species of humans populates Planet Earth today, but our ancestral story is one of riotous diversity. A vivid demonstration of that fact is the story's very first chapter—the early stages of the emergence of primates. During the Miocene epoch from around 23 million to 5 million years ago, hundreds of different apes flourished on many continents. In Eurasia, their fossils span a vast territory from France to China. How did early Miocene apes thrive in such varied landscapes and why did they eventually die out in Eurasia? In their search for clues, many scientists turn to Madagascar, a unique "living laboratory."

Madagascar has been an island since the dinosaur age, and most of its plant and animal species are found nowhere else. The best-known fauna are the agile and charismatic lemurs. With over 100 species, they are the world's most diverse family of primates, ranging from the one-ounce mouse lemur to the *indri* that can weigh twice as much as a house cat. Before humans arrived in Madagascar, there were much bigger ones, including the 350-pound *Archaeoindris*, the size of a male gorilla. These giant extinct lemurs testify to a signature characteristic of primates—their remarkable adaptability.

A case in point was a 1991 discovery in the evocatively named Cave of the Lone Barefoot Stranger in northern Madagascar. While sloshing around in the muddy cave, anthropology student Ted Roese spotted what looked like a stick in the middle of a pool. He pulled out an arm bone, followed by a skull, and called out to fossil experts Elwyn Simons and Laurie Godfrey, who were in an adjoining chamber. They knew at once that the bones belonged to an extinct species they had recently named *Babakotia radofilai*. "We were just floored," Godfrey recalls, "because everything we'd recovered from this species so far had just been fragments, and here was this complete skull." When the team drained the pool, they collected most of the skeleton of what Simons called "one of the most strangely adapted creatures that ever lived."[7]

Babakotia was one of four extinct groups of species known collectively as sloth lemurs. Although related to the indriids, a living lemur group, the anatomy of all four was very different to that of the agile leapers of today's Madagascar forests. Rather than the long, springy legs and short arms of modern lemurs, *Babakotia* had elongated arms, hands, and feet, and short, stocky legs. Together with its curving fingers and toes, these features made *Babakotia* best suited to climbing and hanging from branches rather than leaping from tree to tree. As their name implies, sloth lemurs are thought to have behaved much like South America's slow-moving, tree-hanging mammals.

Babakotia's physical characteristics were taken to an extreme in another sloth lemur species known as *Palaeopropithecus*. Its

(Opposite) The skeleton of *Palaeopropithecus*, one of many species of extinct sloth lemurs that thrived in the forests of Madagascar. Its long arms and hooked hands indicate it was a slow-moving, branch-hanging tree-dweller.

(Opposite) The largest of the extinct sloth lemurs, *Archaeoindris*, was the size of a male gorilla; the 350-pound creature must have moved ponderously and cautiously through the trees. The vast diversity of Madagascar's lemurs, past and present, testifies to the remarkable adaptability of primates.

finger and toe bones (or phalanges) had twice the curvature of a living lemur's. This turned its extremities into hooks, enabling it to move from branch to branch and hang upside down. Nearly every aspect of *Palaeopropithecus* anatomy was specialized for its slow, suspended life in the trees, including a set of ankle bones that paleoanthropologist Bill Jungers described as "simply bizarre...Locomotion on the ground would have been ungainly, perhaps comical, and probably quite rare, except to creep across the ground from one feeding tree to the next when presented with gaps in the forest canopy."[8]

It seems likely that *Palaeopropithecus* was the strange creature described in eyewitness accounts collected by Étienne de Flacourt, the French governor of Madagascar, in the 1650s. His informants spoke of the *tretretretre*, a beast as big as a two-year-old calf with the face of a man, long digits, curly hair, and a short tail. The islanders were said to be very afraid of it and would flee if they sighted it. Other accounts suggest that giant lemurs, not to mention pygmy hippos, may even have survived until just a few centuries ago.

The French were not the first to colonize Madagascar; people from Asia and Africa arrived thousands of years earlier. Cutmarks on the bones of giant elephant birds indicate a human presence around 10,500 years ago. (Some species of these extinct flightless birds were ten feet tall and could weigh up to 1,700 pounds, twice the weight of an average grand piano.) This very early human population may have died out, since there is no subsequent convincing evidence until about 2,400 years ago. Even so, large, slow-moving sloth lemurs were surely easy targets for human hunters. How could they have survived for thousands of years?

This riddle is part of a wider, thorny debate about what drove Madagascar's giant animals into extinction. An authoritative new study argues that for centuries, subsistence hunting had only a minor impact on the island's ecology. Then, around 700 CE, a major influx of new settlers began, leading to a switch from hunting and foraging to herding and farming. Large tracts of forests were gradually cleared for raising crops and pastureland, and the decline in the numbers of the biggest, most vulnerable species started, reaching a climax in around 850 CE. Today, a similar process, fueled by logging, charcoal harvesting, and mining, threatens over 90 percent of Madagascar's lemur species.

Meanwhile, scientists strive to understand the sloth lemurs' vanished world to gain insights into how our primate ancestors expanded into a vast variety of habitats. According to Bill Jungers, they "were part of a much more diverse primate community until 'yesterday.' They document what has been irrevocably lost and provide clues into the factors that continue to drive the extinction event in Madagascar."[9]

THE BLOOD CLOCK

In the outlook of many religions—and of most people before Darwin—humanity occupies a special place in the natural order that is distinct from all other animals. Science began to challenge that view as early as 1699, when British physician Edward Tyson dissected a chimpanzee and noted similarities with humans and other great apes. In the mid-1800s, Darwin and his colleague Thomas Huxley observed anatomical resemblances between chimpanzees, gorillas, and humans, and argued that they must all share a common ancestor. But for over a century afterwards, scientists hoping to fill in details of the great ape and human family trees could only turn to the fossil record, which had many potentially misleading gaps.

Then, in the 1960s, the discovery of the "molecular clock" began a scientific revolution. It was inspired by the ingenious idea that the evolution of biological molecules could provide a measure of how closely different branches of the great apes are related, as well as a timescale for when they split off from each other.

Morris Goodman, a leading developer of the molecular approach, compared a common protein, albumin, which circulates in the blood of humans and primates. His technique involved injecting rabbits with the human version of the protein. This stimulated the rabbit's immune system to make antibodies. Goodman combined the blood he drew from the rabbit with albumin from other primates on a gel plate. By comparing the size of the antibody reactions visible on the plate, he could judge the relative intensity of the immune reactions. The stronger the reaction, the more closely related humans were to the various primates under test.

In 1962, Goodman published a disconcerting conclusion: humans, chimpanzees, and gorillas were all equally related. This cut against decades of work, based on fossils and lingering assumptions of human "specialness," that our origins must be distinct from those of the living great apes and rooted very far back in time, long before the divergence of gorillas and chimpanzees. Now, the new molecular clues placed the human split at roughly the same time as that of gorillas and chimps.

This momentous result was confirmed and refined in 1967 by two University of California scientists, Vincent Sarich and Allan Wilson. They figured out how to quantify the immune response and proposed that blood proteins evolved in primates at a constant rate, slowly changing due to random mutations or "copying mistakes." Like the ticking of a clock, the mutations provided them with a timescale to figure out when two species shared a common ancestor. Sarich and Wilson compared the differences in immune response between the two species, a difference that depended on the length of time it had taken for mutations to accumulate. Then they calibrated the result against evidence from the fossil record to arrive at an estimate of when the species had branched off from one another. Sarich and Wilson's conclusion was even more stunning than Goodman's. They calculated that the molecular differences between humans, chimpanzees, and gorillas had taken a mere 4–5 million years to accumulate—a fraction of the time that experts had previously assumed from studying fossils. Dismissing those assumptions, Wilson said bluntly that if a seemingly human-like fossil was a lot earlier than 5 million years, it could not possibly be our ancestor.

Other molecular techniques emerged that broadly confirmed the findings of Sarich and Wilson, notably the mapping of the human and chimpanzee genomes in the early 2000s. By comparing DNA sequences, again calibrated against the fossil record, geneticists can figure out how different populations and species relate to one another, reconstructing the broad strokes of evolutionary history far back in time. Decoding of the chimp genome

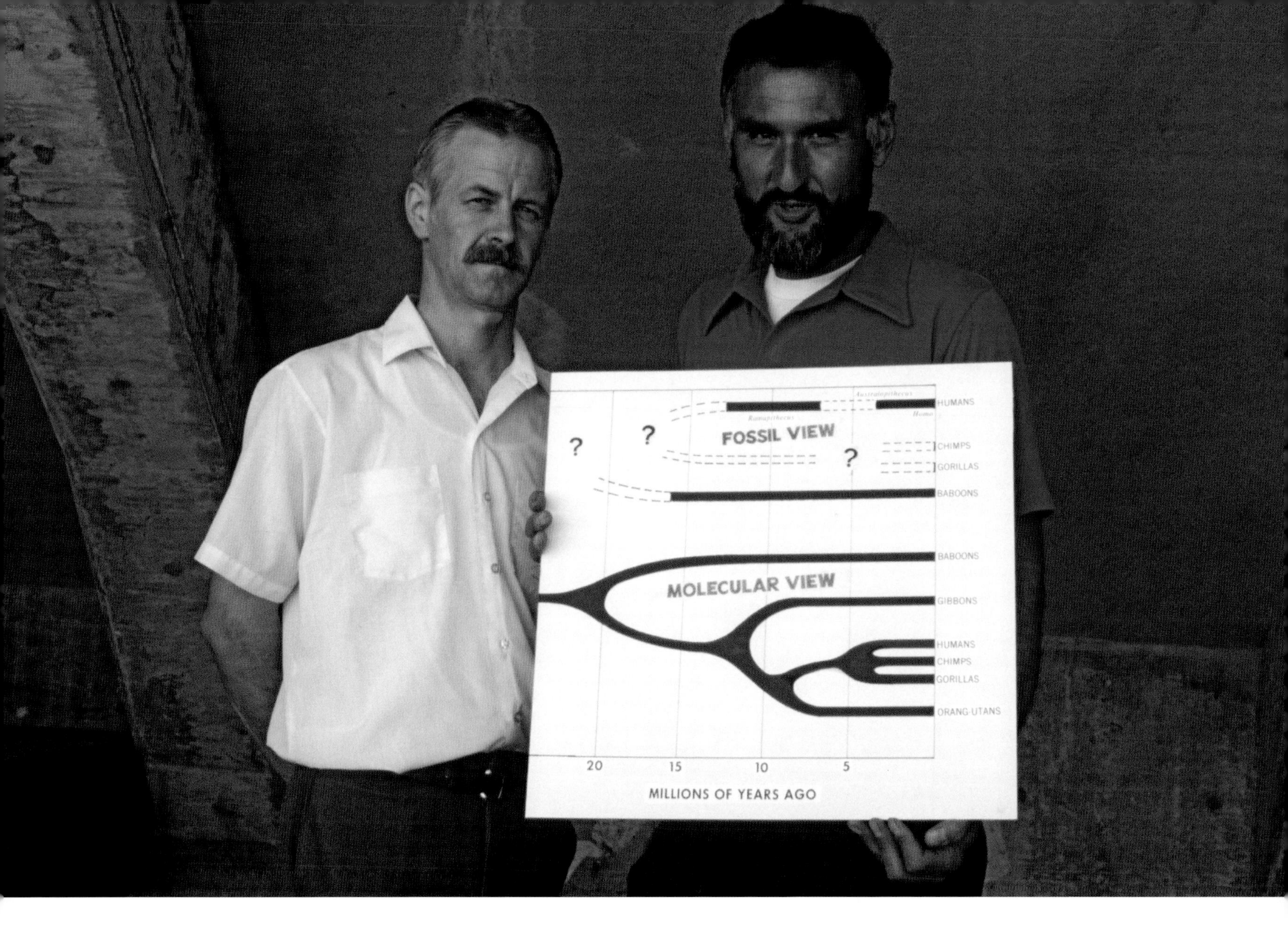

in 2005 captured headlines because it concluded that humans share about 96 percent of our DNA with chimpanzees, making them our closest living relatives. Eventually, enough studies accumulated to sharpen our picture of the split between great apes and humans. It was not, after all, simultaneous; gorillas split off first, followed by the divergence of humans and chimpanzees between 6 and 9 million years ago, according to recent estimates of the genetic and molecular "clocks."

The molecular revolution continues Darwin's and Huxley's quest to understand humanity's place among the primates. Today, we find ourselves classified in a single family, the *Hominidae* or hominins, an identity we share with gorillas and chimpanzees; we are part of the great apes, not separate from them. Our bond with them is more recent and intimate than we ever imagined.

(Above) Berkeley biologists Allan Wilson and Vincent Sarich developed the "molecular clock" that measured the evolutionary distance between humans, monkeys, and the great apes. Their conclusion that humans diverged from them more recently than previously thought was a stunning finding, underscoring humanity's closeness to our gorilla and chimp relatives.

THE GREAT CHIMPANZEE WAR

Striking parallels to human behavior have emerged from Jane Goodall's landmark studies of chimpanzees in Tanzania's Gombe National Park, from their tight social networks and competition for high status to their innovative use of stick tools, passed down over generations. Her first decade of work at Gombe, which began in 1960, led Goodall to view our closest genetic relatives as generally peaceful and benign. "For so many years," she wrote, "I had believed that chimps, while showing many uncanny similarities to humans in many ways were, by and large, rather 'nicer' than us."[10]

Then, in 1971, her familiar chimp world started to unravel. The Gombe community gradually split into two rival factions: to the north, the Kasekela group of eight adult males and 12 females; to the south, the Kahama, consisting of a half-dozen males, three females, and their young. Once tied together by intimate alliances, they now vigilantly patrolled their separate territories, leading to increasingly tense encounters. As primatologist Anne Pusey, then a doctoral student at Gombe, recalls, "We would hear these pant-hoot calls from the south and say to ourselves: the southern males are coming! All the northern ones would go up trees, and there'd be a lot of screaming and displaying."[11] Eventually, the Kasekela males launched stealthy, murderous raids against the southern Kahama community, ambushing lone individuals and inflicting savage beatings. By 1977, they had exterminated almost the entire group and taken over their territory. Many of the attacks that Goodall's team witnessed were so vicious that they gave her nightmares for years afterwards.

What caused this formerly close-knit community to split apart? One theory pointed to the feeding station that Goodall provisioned with bananas so she could observe the chimpanzees at close range. Perhaps this unusual bounty had affected long-standing rivalries, temporarily easing tensions that finally came to the boil again. Or perhaps the intrusion of humans on their shrinking habitat had stressed the chimps to breaking point.

The real answer lay hidden in the decades of daily records of chimp behavior that Goodall's successors have continued to record to this day. Totaling over 350,000 pages, these handwritten notes and check sheets document the lives of hundreds of chimps at Gombe in moment-to-moment, dawn-to-dusk detail. However, they remained relatively inaccessible to researchers until Anne Pusey arranged for them to be digitized in a searchable database—a task, supported by The Leakey Foundation, that has occupied her team for the last 25 years.

To crack the mystery of what triggered the split, Pusey and her colleague Joseph Feldblum recently ran the data through software designed to analyze social networks. This technique

(Opposite) Male chimpanzees in Uganda's Kibale National Park set off on one of their regular territorial patrols. The 190-strong Ngogo group engages in regular "wars" against their neighbors. The motivations behind chimp violence are intensely debated for the light they may throw on the roots of human aggression.

(Opposite) Chimpanzee violence is usually directed toward rivals outside the community and internal killings are rare. An exception was the lethal assault on a top-ranking male named Pimu by four of his underlings, captured on video in 2011 in Mahale Mountains National Park, Tanzania. Pimu was attacked after picking a fight with the second-ranking male.

enabled them to measure how closely each individual was connected to others in the group and how that changed over time. The results allowed them to rule out several theories, including Goodall's banana provisioning. Instead, in a scenario worthy of a soap opera, the analysis pointed to a struggle for dominance between leading alpha male Humphrey and two rivals, brothers Charlie and Hugh. Their once friendly relationships fraying, these top males and their followers increasingly shunned one another and withdrew into separate territories. This coincided with an unusually low number of sexually available females, which may have exacerbated tensions. Eventually, the superior numbers of the northern group meant that the males there could launch surprise lethal attacks on individuals in the south with relatively little risk. The ultimate outcome was the descent into the savage killings that so shocked Goodall.

This dissection of the Gombe "war" is not the only impressively detailed investigation of aggressive conflict between rival chimpanzee groups. For over two decades, also supported by The Leakey Foundation, primatologist John Mitani and his colleagues have tracked an exceptionally large group of up to 200 chimps in Uganda's Kibale National Park. Every few weeks, varying numbers of males in the Ngogo group spontaneously assemble into raiding parties, falling into single file to patrol the borders or invade the territory of their rivals. They creep warily in silence through the forest, behaving as if they are searching out enemies. If they confront larger numbers of opposing chimps, they will usually avoid conflict and bolt for home. But if solitary or just a few enemy males are unlucky enough to stray across their path, they are set upon, bitten, and sometimes beaten to death. While females are generally released, their babies are eaten. "The Ngogo chimpanzees patrol and kill neighbors more frequently than any other chimpanzee group," Mitani says.[12]

The Ngogo data points to similar inducements to warlike behavior as in the case of Gombe. After a decade of raiding involving the killing of 13 rivals, the troop had expanded its territory by 22 percent at the expense of its rivals, and its total population had peaked. In addition, the data shows that by gaining access to new food sources, Ngogo chimps live longer and infants fare better than in other groups. Ultimately, chimp "wars" are battles over resources that enhance fertility and survival.

While there is a risk of exaggerating the violent side of chimpanzee behavior to explain the bloody toll of human wars, the similarities cannot be denied. Looking back sadly on the outbreak at Gombe, Goodall reflected that "suddenly I found that under certain circumstances, they could be just as brutal [as us], that they also had a dark side to their nature. And it hurt."[13]

CHIMPANZEE RECONCILIATION

When he was a 27-year-old doctoral student, Frans de Waal witnessed an incident that would deeply influence his thinking about primates. 1975 marked his first year of observing the world's largest captive chimpanzee colony, consisting of 25 animals on a two-acre island at Arnhem Zoo in the Netherlands. A big fight erupted when a dominant male attacked a female and other apes rushed to her aid. After a pandemonium of screaming and chasing, the group finally calmed down and a tense silence ensued. Many hours later, there was more uproar. Yet this time, the loud chorus of hooting was not over a fight; the animals were watching a male and female chimpanzee embrace and kiss. De Waal was puzzled until it dawned on him that the embracing couple was the same pair that had triggered the earlier fight. "That's when it clicked," he recalls, "and I thought, 'Wow, that's what they were doing!' They were reconciling after the fight and...the whole group understood what was going on—other than me, the scientist. That's how I discovered that chimpanzees reconcile after fights."[14]

This stunned de Waal because at that time, a bleak view prevailed of chimpanzees—and humans—as ruled by instinctual violence. Field studies had begun to reveal the many ways chimp society overlaps with our own, including tool use, intense social competition, and outbreaks of warfare. In 1963, the naturalist Konrad Lorenz argued in an influential book, *On Aggression* that humans had evolved an innate aggression that we had yet to bring under control. A popular image of our early ancestors as "killer apes" began to take hold, memorably dramatized in the opening scene of Stanley Kubrick's 1968 movie *2001: A Space Odyssey.*

As he spent the next six years intently recording the lives of chimps in the Arnhem colony, a radically different view began to take shape in de Waal's mind. Unlike many previous researchers, he combined hard data with his impressions of the relationships, personalities, and power struggles within the group. He discovered that reconciliations after fights are a regular, conspicuous part of chimpanzee social life. Their peacemaking gestures include grooming, kissing, hugging, and hand-holding, all of which greatly lower the probability of renewed conflict between the animals involved. In one of his most influential insights, de Waal gathered evidence that chimps weigh costs and benefits both before they engage in aggressive behavior and in deciding whether to make up afterwards. Among many factors involved are the rank of the two animals involved, the risk of injury, and the value and closeness of their relationship. Ultimately, de Waal came to view aggression and reconciliation as two faces of the same coin—not instincts, but behaviors mostly learned

(Opposite) A hug of reconciliation between two male chimpanzees who had previously been fighting in Uganda's Kibale National Park. Until Frans de Waal's work, many researchers had focused on violence and neglected the importance of peacemaking in chimp society.

in infancy, and essential for expressing and controlling conflict within the chimpanzee's complex social world.

Since de Waal's foundational work, reconciliation has been documented in more than 30 primate species from mountain gorillas to macaques, not to mention in ravens, domesticated dogs, dolphins, hyenas, horses, wolves, and many more animals. The big question is whether the behavior de Waal so meticulously reported at Arnhem is typical of chimpanzees in the forests of Africa. Could the intensity of peacemaking there and at other zoos result from the special conditions of captivity?

So far, only five studies of conflict resolution in chimp populations have been carried out in the wild. In one recent project, then doctoral student Jessica Hartel followed a group of 11 male and six female adults from dawn to dusk through the dense forests and swamps of Kibale National Park in Uganda. During the course of a year, she recorded over 600 bouts of aggressive behavior and 122 episodes of reconciliation. Overall, the chimps in her study reconciled only 14 percent of the time. A similar low rate was reported in the other field studies. In zoo-based research, on the other hand, conflict resolution happens far more often, ranging from 22 percent to 48 percent.

The reason for the discrepancy, Hartel believes, is Kibale's forest setting. Dense undergrowth limits visibility to about 30 feet, so if the conflicting partners end up far apart, the easiest option is simply to quit the scene and disappear. "Why bother to reconcile if you're basically invisible to your opponent, and risk being attacked again?" she says. "Reconciliation is really effective when it works, but renewed aggression happens a lot during conciliatory approaches."[15] If the partners are near each other and particularly if they are close kin or share a valuable friendship, they are much more likely to take the risk and make up. "In the enforced proximity of the zoo," Hartel says, "you're always in sight of your aggressor. Unlike the wild, there's no option to leave or disperse, so reconciliation becomes a more common and possibly more important strategy."

While future research will continue to refine the details, the big picture established by de Waal remains intact: chimpanzees are driven not by built-in hostility but by flexible and resourceful strategies for coping with social conflict. During her year at Kibale, the fiercest fight that Hartel witnessed was resolved by the two partners merely touching their fingertips for a few seconds. "It's really amazing to me," she comments, "how they're able to nonverbally communicate so much about a complicated set of relationships, and reset everything back to where it was with a simple gesture."[16]

(Opposite) Biologist Jessica Hartel approaches one of the chimpanzees from the community that she studied for a year in Kibale National Park, Uganda. Her research marked one of the first in-depth attempts to assess the role of peacemaking in a wild chimp community.

THE FACE IN THE DESERT

Early in the morning of July 19, 2001, four fossil hunters crawled out of their tents into the scorching dunes of the Djurab Desert, which is in the north-central African nation of Chad.[17] It was the last day of the expedition, and their prospects were not encouraging. A sandstorm was brewing, and its choking clouds could half-bury them in their tents for days and force them to don ski masks. They fanned out across a sandstone basin, eyes scanning the ground, watching the hostile landscape not only for fossils, but also for landmines left over from Chad's civil war.

But for the joint French-Chadian team that had explored this wilderness for a decade led by anthropologist Michel Brunet, the stakes were high. By studying the mix of extinct animal species they were collecting, they knew they were unearthing sandstone layers dating to a crucial period around 7 million years ago. The genetic evidence of the "molecular clock" (see p. 60) pointed to this period as close to the split between chimpanzees and humans. Only a handful of possible hominins older than 4 million years had surfaced from the East African Rift Valley and South Africa's limestone caves, the two main centers of research into our ancestral past. Here, thousands of miles away in Chad, Brunet was convinced there was a better chance of finding a very early fossil hominin if the arduous challenges of the fieldwork could be overcome. He hoped it would show a mix of hominin- and ape-like features, the kind of blend that might be expected in a fossil that had recently descended from a common ancestor.

On this July morning, Brunet was in his office in France. Back in Chad, Ahounta Djimdoumalbaye, a young student on the team, spotted the top of a fossil skull close to the edge of a dune. Gently removing it, he turned it over and was surprised to see an ape-like face staring up at him. Apart from its missing lower jaw, it was amazingly complete. Uncertain whether it was a primate or hominin but with a growing sense of its significance, Ahounta shouted to a colleague, "We have what we seek! We are victorious!"[18]

Although badly crushed, the skull of *Sahelanthropus tchadensis*—better known as Toumaï or "hope of life" in the local Goran language—does, indeed, reveal a blend of chimp and human-like features. It contained a chimp-sized brain and it looks like a chimpanzee skull when viewed from the back. From the front, however, its broad, flat lower face resembles that of hominin skulls dated to millions of years later, not the protruding snout of an ape. Its incisor teeth closely resemble a chimp's, but its canines are smaller. While Toumaï's skull has intriguing hints of much later hominins, each of its seemingly advanced features can be matched in other primate species, too. So is Toumaï truly connected with our ancestry? Or is it another type of creature, from a different evolutionary pathway, that ultimately became extinct? If we had other comparable fossil skulls from 7 million years ago, Toumaï's relationship to us might be clearer.

Expert debate about Toumaï's status hinges around perhaps the most distinctive feature of early hominins: bipedalism. Did Toumaï walk upright? Attention has focused on a single skull feature known as the foramen magnum—the opening in the base of the skull where it attaches to the spinal column. In humans, this opening is angled straight down, which allows our heads to balance vertically on our spine; in four-legged apes

(Opposite) The 7-million-year-old fragments of Toumaï, one of the oldest known candidates for a potential human ancestor, lies in the sand of Chad's bleak Djurab Desert shortly after its discovery on July 19, 2001.

like chimps and gorillas, it is angled backwards to accommodate a horizontal spine.

The crushed condition of Toumaï's skull made it hard to be sure of this critical angle. So Brunet took the fossil to the high-energy X-ray facility in Grenoble for high-resolution CT scans. Advanced software allowed researchers to deconstruct and reassemble the fragments into a 3D image of the skull in pristine condition. The result strengthened Brunet's case for a downward-pointing foramen, and therefore the claim that Toumaï did indeed walk upright. Paleoanthropologist Ian Tattersall concluded that the computer model gave "substantial grounds for viewing Toumaï as plausibly—if not definitely—the skull of a biped."[19]

Then, in 2020, the first analysis of a femur found near the skull was finally published. Surprisingly, the researchers found that the leg bone was shaped more like a chimpanzee's than a human's and could not have belonged to a habitual upright walker. So, despite Toumaï's advanced features, its link to our ancestral lineage remains controversial. During the next two million years or more after Toumaï, hominin fossils are so scarce that there is

room for many seemingly ancestral traits, even bipedalism, to have evolved more than once in creatures that have long since vanished. Even if Toumaï is not related to us, it is still a landmark discovery: a first glimpse of the diversity of apes that must have flourished around the time we parted company with our common ancestor. And by persevering with his arduous desert quest, Brunet has shown the potential of other African regions beyond the traditional fossil hunting grounds of East and South Africa. As he puts it, "We have widened the cradle of our origins."[20]

(Opposite) Toumaï's discoverer, student Ahounta Djimdoumalbaye (center), flanked by field technician Fanone Gongdibe and paleontologist Mackaye Hassane Taisso (right) study the reconstructed skull in the lab at Chad's N'Djamena University. (Above) The field team led by Alain Beauvilain collects fossils from the Toumai site on the afternoon of its discovery, when temperatures were reported to have been around 130 degree F at ground level. *Left to right:* Mahamat Adoum, Ahounta Djimdoumalbaye, Fanone Gongdibe.

UNEXPECTED ARDI: A FOSSIL IN FRAGMENTS

On a barren hillslope in the badlands of Ethiopia's Middle Awash region, a pile of rocks commemorates the passing of an ancestor. Imitating a local Afar chief's grave, the cairn marks where one of the oldest creatures to lay claim to human ancestry perished 4.4 million years ago—80,000 generations before the time of Lucy. Reconstructing the first specimen of *Ardipithecus ramidus*, nicknamed "Ardi," took 17 years from the finding of the first fragment until its final publication. It was perhaps the most challenging reconstruction of any single specimen in the human fossil record, and its painstaking analysis confounded many expectations.

In blistering heat near the village of Aramis, a team spent three seasons with their noses pressed a few inches from the ground. "Literally, we crawled every square inch of this locality," says team leader, paleoanthropologist Tim White. "You crawl on your hands and knees, collecting every piece of bone, every piece of wood, every seed, every snail, every scrap. It was 100 percent collection."[21] White calls the dozens of crushed fossil fragments they recovered "road kill[22]... so fragile, you couldn't even breathe on the thing."[23] Ardi's entire skull was so squashed, it was a little over an inch thick. To extract the fragments, the team had to remove the surrounding blocks of sediment, which they then teased apart in the lab under the microscope with dental tools. Meanwhile, in a lab in Tokyo, White's colleague Gen Suwa manipulated 3D CT scans of the skull to create a virtual reconstruction; it took him nine years before he was satisfied with the result. In the end, rarely preserved parts of the skeleton were pieced together, including the delicate hand and foot bones and much of a pelvis.

Why did Ardi's reconstruction merit such Herculean efforts? The period between 4–7 million years ago is almost a blank in the hominin fossil record. Yet it marks the crucial first chapter in the human story, when our lineage split off from those leading to living African apes and we made the transition to upright walking. For much of the last century, scientists assumed that the common ancestor we shared with chimpanzees before the split would have had features that foreshadowed chimp adaptations, such as a pelvis and limbs suited for knuckle walking. Then, as our ancestors began to walk upright, they would have lost those features while the forerunners of today's chimps and gorillas retained them.

(Opposite) A reconstruction of *Ardipithecus ramidus*, a 4.4-million-year-old possible human ancestor. Ardi had the long, apelike forearms and fingers and opposable big toes of a tree climber. Yet the pelvis and other features testify to an efficient upright walker, planting both feet flat on the ground, unlike a knuckle-walking chimp.

The 1994 discovery of *Ardipithecus* led to one of the most challenging studies of any human fossil. The skeleton was in over 100 fragments and its skull was crushed to less than two inches thick. The final reconstruction was a remarkably complete picture of a potential human ancestor more than a million years older than the famous Lucy.

But Ardi turned out to be a highly distinctive creature. At about four feet tall and 110 pounds, she was bigger than Lucy but had a slightly smaller brain than average for Lucy's species. She *did* have some chimp-like traits: long and powerful forearms, long curving fingers, and an opposable big toe, good for grasping branches. Clearly, she was a strong climber at home in the trees. Yet her hands lacked the special adaptations that knuckle-walking chimps have to support the weight of their bodies. Instead, Ardi was a competent upright walker. While chimps lurch clumsily from side-to-side when they walk upright, Ardi's pelvis was adapted so she could walk more efficiently and smoothly. Its wing-like upper part was shaped like our own, providing attachment points for gluteal muscles that helped stabilize her stride. Besides her grasping big toe, her other toes pointed straight ahead and could flex upward at the end of a stride just as our toes do.

Recently, fragments of a new *Ardipithecus* skeleton were discovered at Gona, Ethiopia, during work supported by The Leakey Foundation. According to paleoanthropologist Scott Simpson of the Gona team, the find reveals slight differences in the ankle bone and big toe that suggest an even more human-like gait than the original Aramis remains. Not that she walked entirely as we do today: according to White, "her short legs, long arms, and splayed big toes add up to a creature that is neither human nor chimpanzee, and only accessible through the fossil record." White believes that the long-held "proto-chimp" image of our last common ancestor could be wrong; instead, chimpanzees and gorillas may have evolved into their distinctive forms more recently, well after they split away from us.

But uncertainty over how Ardi relates to the human story is likely to continue. Paleoanthropologist Jeremy DeSilva points out that "fossils do not come with labels, and the closer in time a fossil gets to the common ancestor of humans and the African apes, the more difficult it becomes to differentiate between extinct members of the human, chimpanzee, and gorilla lineages."[24] Only new fossil discoveries of distant ancestors will help resolve the uncertainty; meanwhile, *Ardipithecus* opens up a window on an early stage in our evolution that nobody predicted.

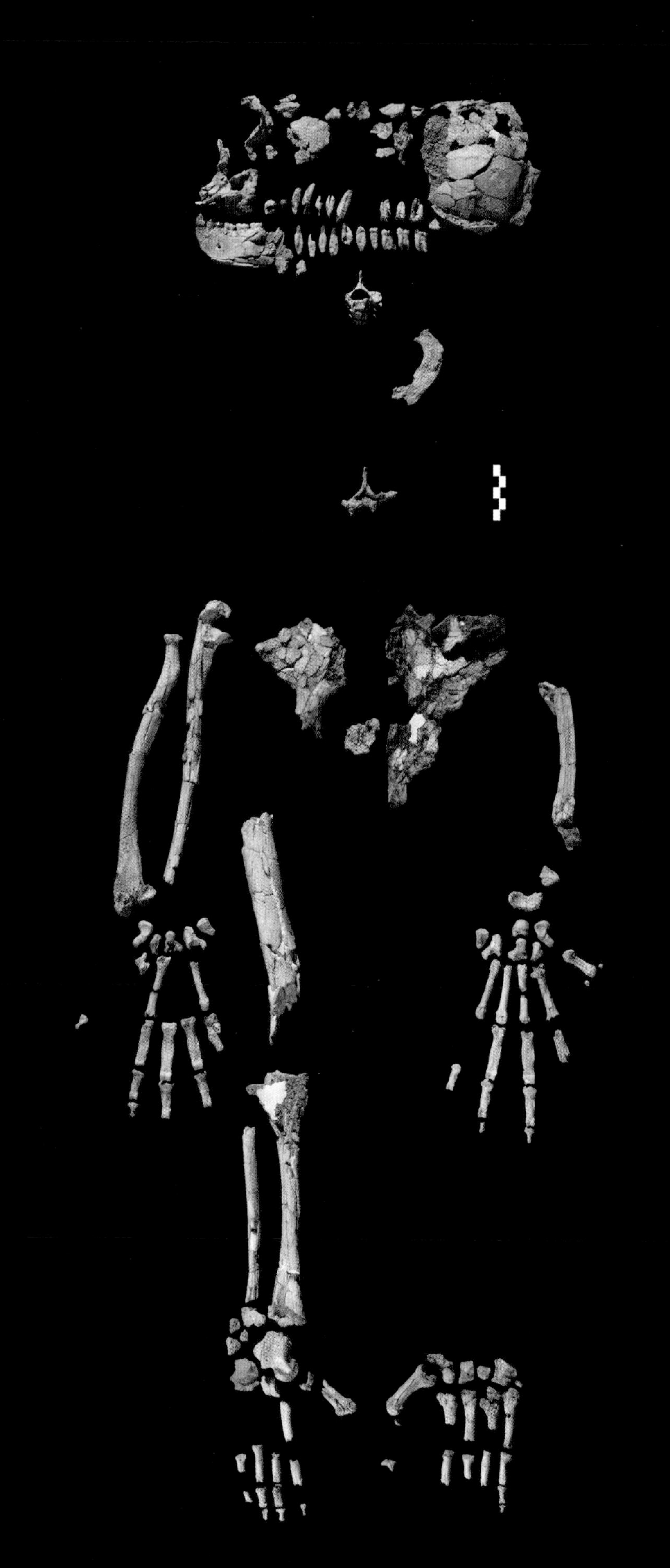

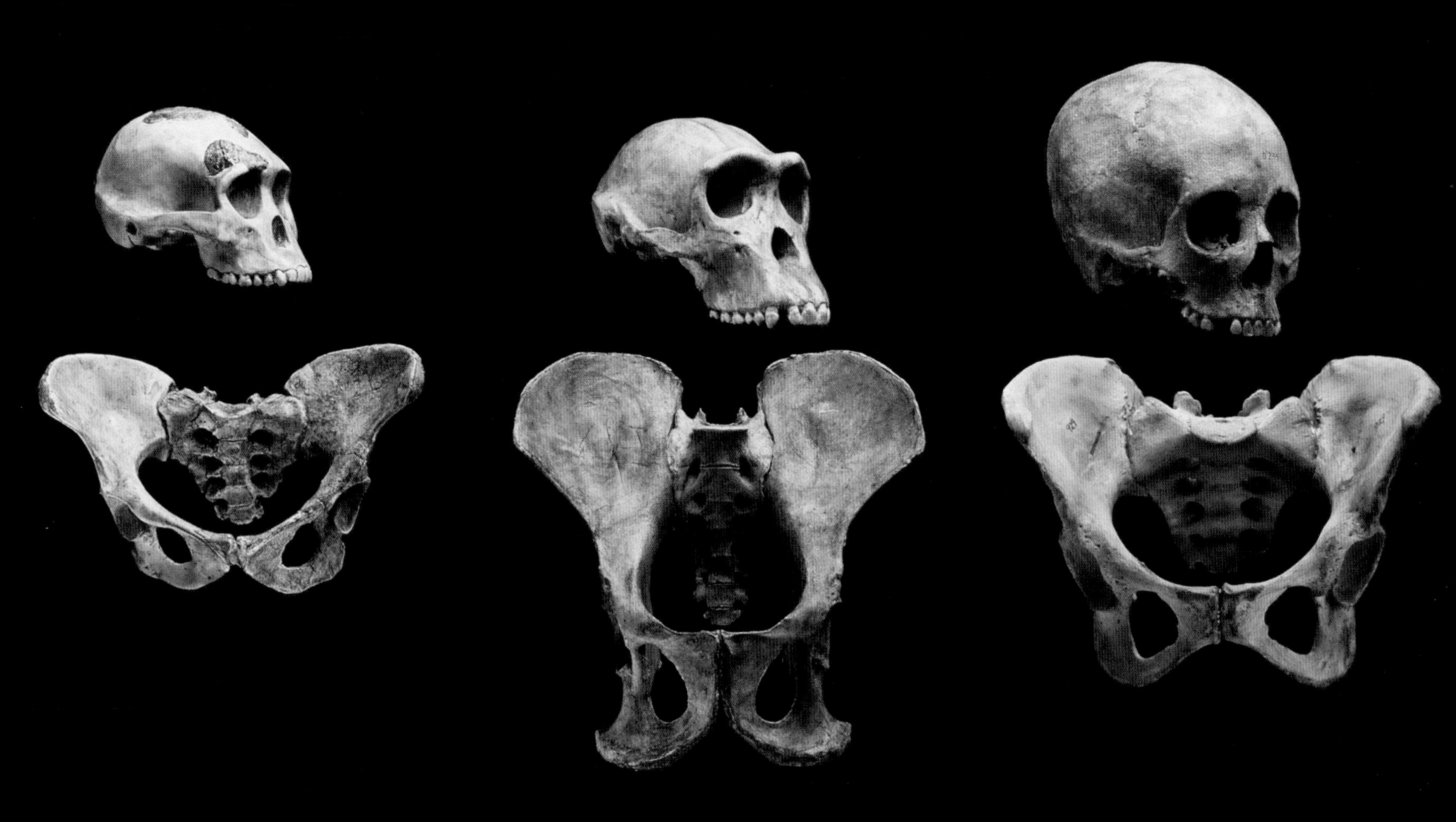

LUCY, THE ICONIC ANCESTOR

What marked our ancestors from other early apes and set them on the evolutionary path leading to *Homo sapiens*? At the start of the 20th century, many assumed that intelligence and a big brain were the answer, and that these crucial advantages emerged at a very early date. But a very different picture began to emerge in 1924 with the first fossil finds of small-brained, upright walkers known as australopithecines in South Africa's limestone caves. It was not until the discovery of the famous Lucy skeleton in 1974 that a complete enough fossil showed what one of those early ancestors, over 3 million years ago, was really like.

On a morning scouting trip in the badlands of Hadar, Ethiopia, the glint of a fossil elbow fragment lying on a rocky slope caught the eye of 31-year-old, newly minted Ph.D., Donald Johanson. Summoning his team in excitement and disbelief, Johanson recovered over a third of the skeleton, including parts of the skull, rib cage, pelvis, and limb bones, which would turn out to be more than 3 million years old. That night, celebrating their success back in camp, the team christened their new fossil, inspired by the Beatles' *Lucy in the Sky with Diamonds* that was playing in the background. The story of her discovery transformed Lucy into a popular icon. "At that time," says Johanson, "it was the oldest, most complete skeleton of a human ancestor. People could visualize an individual when they looked at her."[25]

Subsequent discoveries have amply justified her celebrity. Over the last four decades, more fossils have been found of Lucy's species, *Australopithecus afarensis*, than of any other ancient fossil ancestor. Fieldwork by the Institute of Human Origins has identified over 400 specimens from Hadar alone, representing more than 90 percent of her anatomy. Her species endured in East Africa for at least 800,000 years, nearly three times longer than the timespan of our own species, *Homo sapiens.* Moreover, *A. afarensis* was only one of at least eight other species of australopiths (or australopithecines) that left fossil traces across a vast area, from the desert badlands of Ethiopia to the limestone caves of South Africa. Each one represented a distinctive survival solution to particular habitats and changing climates; some overlapped with each other, and others were still thriving when early *Homo* was foraging on the East African grasslands. But it is *A. afarensis* that has commanded attention as the australopith most widely thought to have given rise to our own species, even though there is no proof of a direct connection.

Despite Lucy's fame and the prolific finds from Hadar and elsewhere, many unresolved questions persist. What sort of a creature was she? Anyone expecting Lucy's brain to be expansive was disappointed. It was grapefruit-sized, similar to a chimpanzee's. She was a little over three feet tall and weighed around 60 pounds, also comparable to a chimp. Her other ape-like features included a snout-like, projecting face and brow ridges, relatively short legs, and powerful, long arms with slightly curving fingers, which made her well-adapted for tree climbing. Spending time in the trees was probably still crucial to the survival of these vulnerable apes, enabling them to find fruit, escape predators, and sleep through the night in relative safety.

Yet much of Lucy's anatomy shows her to have been a committed upright walker. Instead of an ape-like grasping toe and

(Opposite) A reconstruction of the skull and pelvis of Lucy (left) compared to those of a chimpanzee (middle) and modern human (right). Lucy had the projecting snout and brow ridges of a chimp, and her brain was only slightly bigger. But she shared with humans a wider pelvis that curved out to the side and helped to control balance during upright walking, unlike the longer, flatter pelvis of a knuckle-walking chimp.

flexible foot suited for tree climbing, Lucy's stiff foot with its partial arch largely resembled our own, including her big toe that lined up with all the others. As a result, her foot was well equipped to propel her body forward and upward at the end of each springy step. Another striking departure from the apes was Lucy's shorter, wider pelvis, which allowed her muscles to control and balance her body as she took one step after another. While some of these adaptations may have been a handicap in the trees, Lucy's style of walking was an effective formula for crossing open landscapes, perhaps twice as efficient as chimpanzees' habitual knuckle-walking or occasional, upright shuffling gait.

The latest evidence suggests that walking began long before Lucy and the australopiths. It may have started as early as 7 million years ago, since the structure of Toumaï's fossil skull (see p. 71) indicates that it was almost certainly connected to an upright spine. Back then, our ancestors were living in a partly forested environment. Over the next few million years, despite periodic rapid climate swings, there was an overall trend toward a cooler, drier East Africa. This shift in the environment would have affected the distribution of fruit trees, which are the preferred food source of most primates today. During cooler phases, the trees were more spread out across the landscape, putting pressure on them to travel farther and more efficiently to forage and survive. Paleoanthropologist Dan Lieberman concludes that "bipedalism was probably an expedient adaptation for fruit-loving apes to survive better in more open habitats as Africa's climate cooled."[26]

Clear evidence of the resilience of Lucy's species lies in the layers of the Hadar site where she was found, which span some 400,000 years of time. Throughout this era, woodlands supported by high rainfall and mild temperatures alternated with cooler, more arid episodes that favored open grasslands. Despite these shifting conditions, fossil evidence shows that the anatomy of *A. afarensis* remained essentially unchanged, suggesting that they were capable of thriving in very different landscapes.

More tantalizing clues emerge from their outsized jaws and cheek teeth, which are more massive than anything seen in living apes today. Covered in thick enamel and resembling

Lucy's discovery in 1974 created a sensation. For Donald Johanson, seen here shortly after the discovery, the skeleton's visual impact helps explain Lucy's public appeal. "At that time," he says, "it was the oldest, most complete skeleton of a human ancestor. People could visualize an individual when they looked at her."

grindstones, the flat molars of some specimens were twice as big as a chimpanzee's. At first, this rugged anatomy suggested that *A. afarensis* had specialized in crushing tough and abrasive foods like nuts, seeds, and tubers, rather than the soft fruits that most apes prefer whenever they are available. One explanation points to the theory of "fallback foods," the idea that, in times of drought when fruit was scarce, australopiths would have a survival advantage if they could draw on a wider menu than that of other primates. Yet subsequent microscope studies of the wear patterns on *A. afarensis* teeth did *not* reveal the pits and craters expected if they had indeed specialized in grinding up hard, brittle foods; instead, the microscope showed mostly long, wispy scratches, more typical of chewing tropical grasses, sedges, and leaves. This result was confirmed by a follow-up study of the carbon isotopes in her teeth, which again revealed a signature typical of grass and sedge chewers. The paradox remains unresolved, but most experts conclude that australopiths were in all likelihood omnivores, able to draw expediently on a wide variety of foods, including scavenged or hunted meat, as different opportunities, natural settings, and climate conditions allowed.

For Donald Johanson, looking back on four decades of Lucy scholarship and popular renown, she still occupies a special place, "a pivotal point on the human family tree, between older ancestors that are very apelike and more modern, more derived, ancestors, and appears to be the last common ancestor for all the later branches on the human family tree."[27] With the australopiths, our ancestors' earlier experiments in upright walking and dietary resourcefulness were taken to a new level, and would be developed still further by their *Homo* successors. According to anthropologist Carol Ward, the crucial advantages of Lucy's species were "their unique way of moving from place to place on the ground and their ability to make a meal out of most anything they ran across. This gave them the edge over apes that made the difference in the long run, setting the stage for the arrival and evolution of *Homo*."[28]

FIRST FOOTPRINTS

Preserved only in extraordinarily rare circumstances, the footprints of long vanished ancestors are like a snapshot from the past, connecting us with a fleeting moment in the lives of extinct creatures. Paleoanthropologist Marco Cherin recalls the impact of encountering a newly discovered trackway at the world's oldest and most iconic footprint site. "When I was in Laetoli, I could not believe it," he says. "Fossils are completely different from an emotional point of view. [With footprints,] you see the track of the passage of someone. You can read the behavior."[29] Footprints also provide decisive evidence of our early ancestors' crucial transition to upright walking.

In 1974, Mary Leakey was intrigued to hear that fossil teeth had been spotted in a load of building sand collected at Laetoli in northern Tanzania. She went on to recover some thirty hominin specimens at the site. They were nearly identical to Lucy and her fossil cousins from Hadar in Ethiopia, and, at 3.6 million years old, were around the same age. A fossil jaw from Laetoli was chosen as the type specimen identifying Lucy and her species, *Australopithecus afarensis*. While these finds were impressive, nobody was prepared for the astonishing discovery of the trackways.

It started with a playful moment in July 1976, when a group of scientists visiting Laetoli began pelting each other with elephant

Mary Leakey examines the world's oldest footprints at Laetoli, Tanzania, in 1979. Dramatic proof of upright walking, the 3.6-million-year-old tracks were probably made by Lucy's species, *Australopithecus afarensis,* as they walked over freshly fallen volcanic ash.

The best known trackways at Laetoli were made by two individuals with essentially modern feet, featuring well-developed arches and forward-pointing big toes.

dung. During the ensuing battle, two of them dived for cover into a gully where they suddenly noticed strange impressions in a hardened layer of volcanic ash. Over the next two seasons, Mary Leakey's team found over a dozen similar ash exposures and recorded thousands of footprints from monkeys, elephants, rhinos, horses, giraffes, birds, and beetles. There were even splatters from raindrops, which had pelted a fine surface dusting of ash from a nearby erupting volcano. When the rain mixed with the ash, it created a sticky layer that registered the animal impressions. Later, the sun came out and baked the carbonate-rich ash into a hard, cement-like consistency.

Leakey told her team to keep an eye out for hominin prints, but it wasn't until July 1978 that geochemist Paul Abell spotted the first of them. At Site G, the most impressive exposure, two parallel sets of footprints mark the passage of two bipedal creatures heading north. They appear to have been adults, perhaps walking in step with each other, while a third set of prints crossing on and out of their tracks suggests a juvenile tagged along. The prints indicate a strong heel strike and push-off with the toes, motions that closely resembles modern walking. These ancient feet also seem to have had at least a partial arch and a great toe lined up with the others, unlike the angled, grasping big toe of apes.

Tanzanian scientists Fidelis Masao and Elgidius Ichumbaki found a new set of 14 footprints at Laetoli in 2015. Supported by

The Leakey Foundation, Marco Cherin joined them to investigate the site. Just as at Site G, this new trackway at Site S featured a pair of individuals walking north. To judge from the size of the prints and length of the stride, one of the Site S individuals was substantially taller and heavier than any of the others so far recorded at Laetoli. In fact, a wide range of body sizes can be seen in both the footprints and australopithecine fossils as a whole, probably corresponding to a marked size difference between the sexes, or sexual dimorphism. This is characteristic of today's apes like gorillas, orangutans, and baboons, where a few males dominate groups and strongly compete for access to females. However, since there is no way to be sure of the sex and age of the hominins that made the trackways, the idea that they may have followed a more ape-like social life remains controversial.

So which creature made the Laetoli tracks? Since the only hominin fossils found at Laetoli belong to Lucy's species, the most likely answer is *Australopithecus afarensis.* Much debate has centered on subtle features of the prints, which some experts interpret as signaling that the Laetoli hominins had not yet attained a fully modern gait. Others, however, say the tracks prove that a fully developed, essentially modern form of walking had already emerged more than 3 million years ago.

To try to settle the issue, several ingenious experiments have been staged that compare the Laetoli tracks to the footprints made by humans and chimpanzees. One such trial involved a group of "habitually barefoot and minimally shod" volunteers from the northern Lake Turkana area in Kenya, who walked across a trackway of ancient sediment similar to the soil at Laetoli. These impressions were then compared to ones created by upright-walking chimps in a primate locomotion lab in the U.S. The experimenters conclude that the Laetoli footprints register the motion of a creature with a well-developed style of bipedalism that was nonetheless "slightly but significantly different from our own"[30]—specifically, a more chimp-like flexing of its lower limbs. In the experimenters' view, the Laetoli prints testify that the evolution of walking was still a work in progress.

However the details are interpreted, no one questions the huge significance of the Laetoli footprints; they confirm the already strong fossil evidence that Lucy and her kind were habitual and proficient upright walkers. Writing evocatively of the original discovery, Don Johanson notes that "partway along the trail, the hominids appear to have paused, turned, and looked westward. What caught their attention? The answer is lost in time, but after the pause they resumed walking in the original direction—as if they knew exactly where they were going."[31]

THE PEACE CHILD

In December 2000, Ethiopian scientist Zeresenay "Zeray" Alemseged ventured into a remote corner of the Ethiopian Rift Valley. Dikika is a lunar-like landscape, blasted by hot winds and riven by shoot-outs between pastoral tribes contending for land and water. Despite the risks, Zeray was determined to explore its untouched scientific potential. On the afternoon of December 10, "we decided to survey this hillside," he told NOVA viewers, "and the first thing we spotted was a cheekbone of the face."[32] Peering up from the slope, the tiny face had a smooth brow and short canine teeth that immediately flagged it as a hominin. What emerged from the rock was the world's earliest child, and she has given us unique insights into the development of an ancestral infant. Zeray named her "Selam," or "Peace" in the local language, to express his hopes for the conflict-torn region.

Zeray had recently earned his Ph.D. at the University of Paris and the Muséum National d'Histoire Naturelle in Paris. Now he was back in his homeland, the first Ethiopian ever to lead an Ethiopian field project. Partly funded by the Institute of Human Origins and partly out of Zeray's own pocket, he says, "I was the driver, so I didn't need to pay a driver; I was the cook, so I didn't need to pay a cook; and I was the only scientist."[33] Three field assistants made up his shoestring team. They took the sandstone block that encased the baby's skeleton to the National Museum in Addis Ababa, and began an arduous dissection of the fossil with dental tools. "We spent hours and hours and years and years, and I removed the sand grains, grain by grain, working every day."[34] It took six years before Zeray was ready to publish the fossil and the intricate work of cleaning it continues today.

An underlying layer of volcanic ash dates Selam to 3.3 million years ago, some 150,000 years older than Lucy, the adult female *Australopithecus afarensis* that Donald Johanson discovered in the neighboring Hadar region in 1974. Nevertheless, Selam is evidently the same sex and species as Lucy. CT scan images of the growth of her unerupted adult teeth hidden inside her jaw reveal that she was only about two and a half at the time of her death. A geologist on Zeray's team believes that a flash flood along the nearby Awash River may have swept her away and swiftly sealed her remains in mud.

Selam's melon-sized skeleton, curled up in a crouching position, is much more complete than Lucy's. Tucked under her skull is an almost intact spinal column with ribs and shoulder blades still in place. There are tiny curving fingers, nearly an entire foot, and pieces of limb and leg bones, including kneecaps the size of macadamia nuts. There is even a hyoid, a little horseshoe-shaped bone in the throat that helps support the voice box, and is so fragile that only two others are known in the fossil

(Opposite) Selam, the world's earliest child, belonged to Lucy's species, *Australopithecus afarensis*, but lived some 150,000 years before Lucy. Barely three years old, she was fully capable of upright walking.

(Opposite) Selam's exquisitely preserved skull required years of lab work to reconstruct and revealed unique insights into the growth and development of Lucy's species.

record. Intriguingly, Selam's hyoid is more chimpanzee-like than human. In chimps, the bone supports inflatable air sacs that are thought to help them make rapid sequences of hooting calls without hyperventilating. Losing the air sacs may have been an essential prelude for our ancestors to develop a voice box capable of complex, articulated speech.

All these rare clues tell an intriguing story. Below the waist, Selam has advanced features such as leg bones and a broad heel that would have made her a proficient upright walker. Selam's uniquely preserved spine reveals another such adaptation. Compared to a chimpanzee's, it has fewer rib-supporting vertebrae and more vertebrae in the lower back, a structural shift more suited for efficient walking and running. But her long curving fingers, limber big toe, and gorilla-like shoulder blades suggest that she still sought shelter in the trees, perhaps sleeping there to escape predators at night while walking upright during the day.

Selam's almost intact skull allowed Zeray's team to calculate that her brain size was around 330 cc, roughly the same volume as a large grapefruit or the brain of a small, three-year-old chimpanzee. By that age, most chimp brains are fully formed. But, to judge from other australopith fossils, their adult brains typically reached between 400–550 cc. Zeray's team therefore concluded that Selam's still had another 25 percent or so left to grow. She had been growing faster than a modern infant yet slower than the rapid early growth spurt of a chimp. It looks as if the highly distinctive lengthy period of human childhood compared to the great apes had already begun—a pattern necessary for our ancestors to develop their increasingly complex social skills and survival strategies.

Nearly 20 years after Selam's discovery, Zeray's team continues to tease new clues from the exceptionally preserved skeleton. Meanwhile, his field project, begun on a shoestring, has now grown into the Dikika Research Project, an ambitious international collaboration of scientists that returns to Ethiopia each year to search for new evidence of Selam's long vanished world.

A TALE OF TWO NUTCRACKERS

The Cradle of Humankind, a UNESCO World Heritage site in the rolling grasslands of the high veld country about an hour's drive from Johannesburg, is home to many of South Africa's most spectacular fossil discoveries. In 1938, a local schoolboy took Scottish anthropologist Robert Broom to the remnants of a cave known as Kromdraai to show him where he had picked up a bizarre skull. It housed a tiny brain and was a formidable chewing machine, with thickly enameled molars in a massive jaw. To support the huge muscles attached to the jaw, its cheek bones flared out on both sides, giving it a wide, saucer-shaped face. Broom named the new species *Paranthropus robustus*; *Paranthropus* meant "beside man," reflecting Broom's conviction that it had thrived alongside early members of the *Homo* family for a million years, then died out. Its outsized jaw and teeth indicate a very different diet and way of life from *Homo* and are clues to why it ultimately became extinct.

In 1959, these remarkable South African finds were overshadowed by the worldwide publicity attending Louis and Mary Leakey's discoveries in Tanzania's Olduvai Gorge. After decades of searching for humanity's roots in East Africa, Louis triumphantly proclaimed that a robust skull found by Mary, which he called *Zinjanthropus boisei*, was the earliest human toolmaker. Two years later, the Leakeys began finding daintier fossils, which they identified as specimens of early *Homo*, and so they concluded that *Zinj*, like its southern counterpart *P. robustus*, belonged to an extinct side branch of the human family. Today, this rugged species is classified as *Paranthropus boisei*. It has even more extreme chewing adaptations than *P. robustus*. The flat surface of its first molar is nearly as big as a dime and almost twice the size of our own. On first glimpsing *Zinj*'s mighty teeth, prehistorian Phillip Tobias is said to have exclaimed, "I have never seen a more remarkable set of nutcrackers,"[35] inspiring *boisei*'s nickname, "Nutcracker Man."

Just over thirty years later, in 1992, a retired South African geologist called André Keyser was looking in the hills near Kromdraai for an ancient site to excavate as a hobby. After the owner of Drimolen farm invited him to explore his property, Keyser spotted an abandoned mining pit, crawled under some trees, and immediately discovered baboon and elephant fossils next to a cave entrance. Less than a week later, he had found the tooth of a hominin—the first of nearly 80 specimens from what would prove to be one of South Africa's richest fossil sites, still under excavation today. "My hobby exploded on me," Keyser said.[36] In October 1994, a volunteer at Drimolen unearthed part of an upper jaw containing two molars. Keyser took over, brushing the soft sediment aside to reveal a skull that had been

(Opposite) Nicknamed Eurydice, the most complete skull ever found of *Paranthropus robustus*, was found alongside the massive jaw of Orpheus (to its left) at South Africa's Drimolen Cave in 1994.

(Opposite) The reconstructed skull of "Zinj," the Leakeys most famous find at Olduvai Gorge in East Africa. It belonged to *Australopithecus boisei*, a side branch of the human family. Like *Paranthropus robustus* in South Africa, *P. boisei* lived alongside more advanced ancestors before finally becoming extinct.

invaded by plant roots and an ant colony. Over five years, its 2-million-year-old fragments were pieced together to form a stunningly complete fossil of *Paranthropus robustus*.

Nearly a thousand miles separate the heavy-jawed "nutcracker men" of South and East Africa. Did they belong to a single species, or two very closely related ones? And why had they evolved such extreme jaws and teeth? In 1954, anthropologist John Robinson, who started his career as Broom's assistant, came up with his influential "dietary hypothesis." In this scenario, the millstone-like molars of robust australopithecines had adapted to grind tough, fibrous plants, seeds, and nuts. Meanwhile, the lightly built *Homo* pursued a more varied diet that included fruits and meat. As the climate grew increasingly arid toward 1 million years ago, the narrow specialists lost out while the smarter, more versatile *Homo* survived. So, when a leading expert on ancient tooth wear, Peter Ungar, first investigated the teeth of *boisei* in 2008, he was shocked. He was expecting to see tooth surfaces pitted with lunar-like craters, the signature of a habitual eater of hard, brittle foods like nuts and seeds. But under the microscope he saw only wispy scratches, which are typically created by chewing soft grasses. He later confirmed the evidence by comparing isotope ratios in *boisei*'s tooth enamel, which showed that over three-quarters of its diet consisted of grasses. The verdict was clear: "Nutcracker Man" had not cracked nuts! Instead, it had used its massive jaws and teeth to grind grass like a cow, an adaptation seen nowhere in living primates today.

Equally surprising, the tooth surfaces of South Africa's *P. robustus* revealed a different eating strategy. "Just by looking at the similar shapes of the skulls and teeth in the two species," Ungar says, "you would be led to believe they ate the same things."[37] But under the microscope, he saw a mixture of pitting and scratches, indicating that *P. robustus* had followed a broader, versatile diet. They had probably dined on fruits and nuts whenever they were available, and fallen back on grasses and sedges in leaner times. Since the ability to chew "fallback foods" would help them survive a severe drought, evolution favored the emergence of rugged jaws and outsize teeth.

Why did both these feeding strategies eventually fail? Reconstructions of the earth's ancient climate around 1 million years ago suggest abrupt shifts between wet and dry conditions. While the "nutcrackers" of Drimolen and Olduvai could have got through occasional times of scarcity, at some point the wild swings in their environment proved too much, and their populations dwindled—or so the theory goes. During their heyday, however, these mighty-jawed creatures were more than a mere evolutionary "dead end:" they thrived for at least five times longer than we, modern humans, have so far.

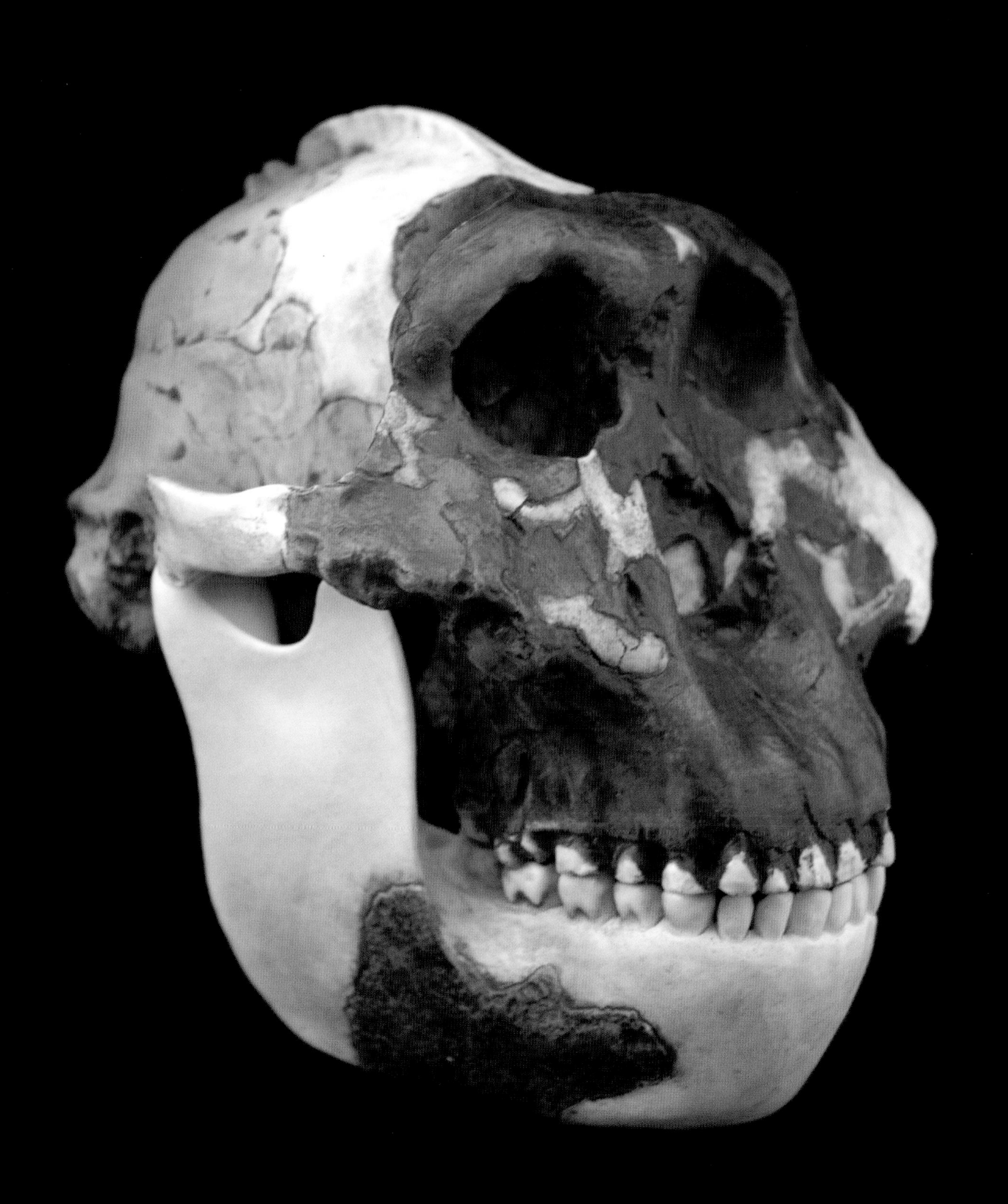

SEARCHING FOR THE EARLIEST TOOLS

At first glance they might look like broken garden rocks. But to scientists, the sharp-edged river cobbles unearthed at Gona—near Hadar, where Donald Johanson discovered Lucy—are full of significance. Among humanity's earliest tools, they have attracted intensive investigation for over two decades.

The Gona project was founded by Ethiopian-born paleoanthropologist Sileshi Semaw. Raised in a deprived neighborhood of Addis Ababa, Semaw had to overcome many challenges until a scholarship enabled him to pursue graduate studies at Rutgers University. In 1992, he carried out the first systematic excavations at Gona in the harsh badlands of Ethiopia's Awash basin. When eventually published, two of these sites triggered worldwide headlines; they consisted of thousands of undisturbed stone tools securely dated to around 2.6 million years. "That was extraordinary," says Semaw, "because it was the first time that abundant materials dated to such antiquity were discovered in the world."[38]

In 2000, Semaw's colleague and co-director of the Gona project, Michael Rogers, made another momentous discovery. "You needed to be a goat" Rogers jokes,[39] recalling how he scrambled around Gona's steep, crumbling gullies on that fiercely hot February morning. At a spot they later called OGS-7, Rogers glimpsed an unusual scatter of sharp little stone flakes. They

(Opposite) Paleoanthropologist Sileshi Semaw holds a 2.6-million-year-old river cobble with deliberately flaked sharp edges. At Gona in Ethiopia, Semaw co-led a team that discovered the oldest stone tools directly associated with butchered animal bones.

(Previous pages) Gona in the Afar region, Ethiopia, is a harsh landscape today, but two million years ago, it was a mosaic of grasslands and forests fringing the lakes and tributaries of the ancient Awash river.

were eroding from a four-inch-thick layer of volcanic ash, mixed up with fragments of animal bone. When Rogers's team dug down to expose the layer, they revealed hundreds more tool and animal bone fragments in near-perfect condition, lying undisturbed for more than 2 million years. "We kept pinching ourselves," he says. "It looked as if the tools were made yesterday."[40] Moreover, at other nearby sites, the animal bones mixed in with the tools bore cut marks—telltale signs of butchering. They had found the world's earliest stone tools definitively associated with processing meat.

OGS-7 yielded many other intriguing clues. Whoever fashioned the tools had fetched cobbles from a streambed, then struck them together systematically to produce sharp-edged flakes capable of cutting open a carcass. After studying gravel from those ancient streambeds, the Gona researchers realized that the tool makers had carefully selected only relatively rare, fine-grained volcanic cobbles, which were easier to work with and produced a sharper edge. Not only had they picked out the best kind of stone for flaking, but modern experimenters have found that it takes careful hand-eye coordination and many hours of practice to reproduce the skills apparent in the ancient debris. The toolmaker had evidently held a riverbed cobble in one hand while systematically striking sharp flakes off it with a hammerstone grasped in the other. "If this was the earliest site in the world," says Rogers about OGS-7, "we expected things to be crude, but the tools appear to have been well made."[41] Another site, recently discovered at Bokol Dora in Ethiopia, has similar relatively well-crafted tools that now push the date of this innovative industry back to before 2.6 million years. That begs the question, just how much earlier did the very first, cruder efforts at toolmaking begin?

On July 9, 2011, archaeologist Sonia Harmand was driving her jeep in a desolate area known as Lomekwi near the shores of Lake Turkana in Kenya, when she took a wrong turn and ended up in a dead-end gully. Deciding to take a look around, she stumbled on a scatter of stone debris. Her team eventually recovered around 150 artifacts that are larger and less well-shaped than the Gona tools. Some flakes are as big as a human hand while other worked rocks weigh up to 30 pounds—seemingly too unwieldy for butchering meat. Experiments suggest that the Lomekwi tools were the end product of a more primitive stone-working technique than at Gona or Bokol Dora. Instead of holding a cobble in one hand and carefully striking blows with a hammerstone as at those sites, the Lomekwi toolmakers used two different methods: they either held a cobble in both hands and then struck it against a larger rock or anvil, or placed a cobble on the anvil and smashed it with a hammer stone. Today, chimpanzees in the Ivory Coast's Taï forest in West Africa use a similar hammering technique to crack open hard-shelled nuts against an anvil stone (see p. 100).

Dated to 3.3 million years old, the Lomekwi site represents a singular, provocative discovery; no other clear evidence of toolmaking has yet turned up in the 700,000-year gap before Bokol Dora. Does it represent a one-off innovative episode? Or a transition between using natural rocks, as chimps sometimes do when they crack nuts, and the first deliberate crafting of sharp-edged stone tools? And who was banging the rocks together? Lomekwi is roughly contemporary with Lucy and well before the earliest known *Homo.* That raises the possibility that australopithecines may have been the first toolmakers. If so, then their relatively small brains, on average only slightly bigger than a chimp's, were evidently no obstacle to setting our ancestors off on the long path that has led to today's technological civilization.

CHIMPANZEES OF THE TAÏ FOREST

In 1979, Christophe Boesch and Hedwige Boesch-Achermann launched the first investigation of chimpanzees in the tropical lowland rainforests of West Africa—a drastic contrast to the drier, more open woodland of East Africa, where Jane Goodall made her celebrated studies at Gombe in Tanzania. Venturing into the Taï National Forest in the Côte d'Ivoire, the Swiss husband-and-wife team spent five years habituating a single community of just over 100 chimpanzees to their presence. Each day, they had to figure out how to locate and follow them through the dense maze of tropical vegetation. Over 16 years of patient observations supported by The Leakey Foundation, they recorded very different behavior than Goodall had seen in East Africa.

Female chimps in the Taï forest are expert nut crackers and tool users, and they devote hours each day to opening hard-shelled nuts. Remarkably, they create workshops around the bases of trees, assembling collections of wooden and stone hammers and using rocks or tree roots as anvils. Each type of nut calls for a different type of hammer; wooden ones for soft Coula nuts, and, for tough Panda nuts, granite rocks, weighing up to 45 pounds. The chimps carefully deliberate before choosing the right hammer for the job, and wield it with great precision. "Time and again," the Boesch-Achermanns noted, "we have been impressed to see a chimpanzee raise a twenty-pound stone above its head, strike a nut with ten or more powerful blows, and then, using the same hammer, switch to delicate little taps from a height of only four inches."[42] While no one has ever observed chimps flaking stones, many of the actions involved in nut-cracking foreshadow the intentional crafting of stone tools—from carefully selecting the best hammers to precisely aiming percussive blows.

Nut-cracking expertise in the Taï forest is a learned behavior, passed on from mother to infants through years of teaching and imitation. Neighboring chimp groups have distinct preferences for the type of hammers they use, while others in the same forest less than 20 miles away have no nut-cracking tradition at all. Despite the abundance of hard-shelled nuts in East Africa, nut cracking has never been observed in Gombe or elsewhere in the Rift Valley.

The Boesch-Achermanns reported another striking specialty at Taï: cooperative hunting. As in East Africa, hunting and sharing meat are vital for cementing social status and alliances. But they observed the Taï chimps hunting far more frequently and collaboratively than anywhere else where hunting has been studied in detail, including Gombe. During the birth season of the chimps' favorite prey, red colobus monkeys, hunting at Taï is a daily activity. Stealthily encircling a group of colobines, the hunters close in, taking on specific roles as blockers or attackers as the monkeys flee in panic. This level of close cooperation may be due to Taï's thick rainforest, where visibility is limited to only fifty feet or so. In the more open woodland at Gombe, the group takes a more scattershot approach, with individual hunters pursuing prey in many directions after the ambush.

The Boesch-Achermanns suggest that the demands of "3D arboreal hunting" in the tropical canopy may have stimulated more complex survival strategies, such as enhanced social cohesion and cooperation, in rainforest chimpanzees compared to those that evolved in the Rift Valley. It is a provocative idea, and a new generation of field researchers at Taï is engaged in testing and confirming the unique aspects of behavior observed there. Their work is a race against time in the face of drastic drops in West African chimp populations due to poaching and disease, but the efforts of the Wild Chimpanzee Foundation, founded by the Boesch-Achermanns, offer hope for their survival.

(Opposite) Female chimpanzee Pola cracks Panda nuts with a stone hammer while her infant daughter, Placali, watches closely in the Taï National Forest, Côte d'Ivoire, West Africa.

CHIMPANZEE HUNTING

At first, primatologist Jill Pruetz could hardly believe her research assistant's report. He had watched Tumbo, a female chimpanzee, strip the leaves off a small branch, trim one end to make a sharp point, and then use it like a spear to poke at a bush baby—a pocket-sized nocturnal primate—that lay sleeping in its nest in a hollow tree. When Tumbo finally captured the bush baby, she bit its head off.

Southeastern Senegal lies at the far western extreme of chimpanzees' range in Africa. The landscape at Fongoli is mostly open, semi-arid grasslands with a few scattered trees. In 2001, with Leakey Foundation support, Jill Pruetz set up her study of a chimpanzee band in this harsh bushland, which teems with ticks, cobras, and black mambas. A lack of creature comforts is no obstacle to her. She has suffered a bout of tick fever and dug a botfly out of her foot, and she comes down with malaria nearly every year.

As Pruetz followed the Fongoli chimps around, she began noticing their unusual behavior. Unlike most other chimp populations, they spend much of their time on the ground in a single, large communal group. Most chimps are afraid of water, but at the peak of the summer heat, when temperatures often soar above 100° F, they frequently soak in pools to cool off or take shelter in nearby caves. During a period of full moon,

Primatologist Jill Pruetz holds a stick used by a female chimpanzee called Lucille to hunt for bush babies in their tree nests at Fongoli, Senegal, West Africa.

(Opposite) At Fongoli, a female chimp called Eva, watched by Tess, spears a bush baby in its nest. Fongoli is the only chimpanzee group reported to use tools regularly as hunting weapons.

they move around and forage at night. All these are remarkable observations, but the report of Tumbo's spear hunting was stunning news. No one had ever seen chimpanzees hunt with tools before, and in most other communities, dominant males, not females, lead the hunt.

It soon became clear that the observation was no fluke. In a 2015 study, Pruetz's team documented more than 300 cases of bush baby spear hunting. Females carried out more than half of these hunts and were the most avid tool makers and users. When males *did* participate, they tended to wait for opportunistic moments, chasing down fleeing bush babies after they had been flushed out of their nests by females or juveniles. Even more striking, adult males tended to respect the ownership of the meat acquired by females, juveniles, or low-ranking males. In other chimp populations, dominant males typically steal around a quarter of the kills made by others; at Fongoli, this happened in only 5 percent of the cases. Why is there more social cohesion and tolerance at Fongoli than elsewhere?

Part of the answer is that, overall, males are still the most successful hunters. As well as bush babies, they chase down other prey such as vervet monkeys and baboons that females, who usually have infants clinging to them, have little chance of catching. The tree-bound bush babies, on the other hand, provide meat that females and young chimps can readily acquire with the aid of tools. Pruetz believes that another explanation lies in the relative sparseness of the semi-arid grasslands compared to the rainforest. "In the savanna, food sources are few and far between, which can exert pressure to make more sophisticated tools and develop more innovative hunting behaviors. We see it today with the chimps at my site, and hominins may have responded to this challenging habitat in the same way."[43]

The transition in East Africa around 2 million years ago, from a forested environment to a savanna resembling Fongoli today, is often held responsible for accelerating the pace of our ancestors' tool use, social development, and growing brain. Pruetz's groundbreaking work at Fongoli points to the possibility that female inventiveness may have played a key role along the winding path that led to our humanity.

BONOBOS' POWERFUL SISTERHOOD

The first time primatologist Frans de Waal saw bonobos was in a Dutch zoo. Then a student, de Waal was struck by how differently they looked and behaved compared to chimpanzees. "Chimps are brawny bodybuilders," he says, "whereas these apes looked rather intellectual. With their thin necks and piano-player hands, they seemed to belong in the library rather than the gym."[44] Then he witnessed something strange. A male and female were squabbling over a cardboard box when they abruptly stopped fighting and began mating. Dismissing it as an anomaly, it was only when de Waal began intensive videotaping of bonobos at the San Diego Zoo that he realized how sex permeates every aspect of their social life. The more you watch bonobos, de Waal says, "the more sex begins to look like checking your email, blowing your nose, or saying hello. A routine activity."[45] The zoo studies carried out by de Waal and his colleague Amy Parish showed how constant sex helps defuse rivalries, aggressive behavior, and competition over food among bonobos. In fact, sexual activity between females is a crucial key to the unique structure of bonobo society, which breaks the evolutionary rules governing most other species. Understanding the riddle of bonobo female power has attracted a fresh generation of researchers who are gradually unlocking its secrets.

(Opposite) Female bonobos socialize at the Wamba reserve in the Democratic Republic of the Congo.

Among them is Leakey grantee Liza Moscovice. Since 2011, she has been based at the LuiKotale site in the Congo, reached by a two-hour flight, a 15-mile jungle hike, and a dugout crossing of the Lokoro river. Here, local communities collaborate with international researchers to monitor and protect several bonobo groups. Moscovice's work at LuiKotale shows that female solidarity is the most striking feature of bonobo society. Like chimps, male bonobos remain in their birth community and females migrate to join other groups when they reach adolescence. But chimp bands are ruled by dominant males who bond together in strong coalitions while, at LuiKotale, male bonobos remain attached to their mothers. It is the females who form a wide range of strong friendships and alliances within the groups they migrate to, even though they are not genetically related. "Females prefer to cooperate with other unrelated females and we don't quite know why yet," says Moscovice. "Sometimes, they'll even cooperate at the expense of their kin; they'll avoid helping their adult sons if it will cause conflicts with their female friends. This is exceptional among primates—basically, putting friends above your own family."[46]

This "secondary sisterhood," as de Waal calls it, is the strongest force in the bonobo world. Females hold the top alpha position and cement their same-sex alliances with constant sexual activity—what primatologists term "genito-genital" or "G-G" rubbing. This physical contact seems to be how females signal trust as they form temporary alliances to deal with conflicts that erupt within the group. "It's *not* that the high-ranking females cooperate against the low-ranking females," notes Moscovice. "It's more common that low-ranking females get help from high-ranking females against males that they're in conflict with."[47]

Belligerent or bothersome males are put in their place in no uncertain manner. In one case observed at Wamba, four males focused unwanted sexual attention on a low-ranking female, triggering sudden retaliation from three high-ranking females, who piled onto one of the offenders, biting and screaming. Finally, "he dropped from the tree and ran away, and he didn't appear again for about three weeks," says primatologist Nahoko Tokuyama, who witnessed the attack. "Being hated by females is a big matter for male bonobos," she adds.[48] One explanation for the strength of female bonobo bonds is that they help to defuse male aggression and keep infants safe.

Remarkably, female social networks can extend beyond their own local group to other bonobo communities. Unlike the violence that often flares up between highly territorial chimps, bonobo communities occasionally coalesce and mingle, with generally peaceable results. In her last field season, Moscovice observed a fusing of two groups, totaling close to 80 individuals, that lasted a month and a half. "They weren't always together, they were splitting up and coming together again just like a regular community. There was sharing of fruit trees and some serious aggression, so it wasn't all peaceful. But there was a lot of tolerance, especially among the females."[49] During this extended visit, unrelated females from the two groups would express their long-term friendships through mutual G-G rubbing, grooming, and feeding together.

One of Moscovice's main goals is to understand how these female friendships work and the benefits they bring. In urine samples collected before and after G-G rubbing within communities, she has discovered high levels of oxytocin—a hormone well-known in the human world for encouraging feelings of trust and empathy in social encounters. In much-publicized human experiments, a mere whiff of oxytocin, delivered to the brain via a nasal spray, was usually enough to make participants more trusting of strangers. Although the effect in humans has proven to be weaker than once assumed, the high oxytocin levels detected in female bonobos after physical contact may partly explain what motivates them to cooperate with so many female partners.

Writing over two decades ago, Frans de Waal was tempted "to speculate that bonobos are flooded with oxytocin...[which] might explain their high levels of social and sexual bonding."[50] Moscovice's research is helping to confirm de Waal's speculation, but the role of oxytocin is only part of the story. "If bonobos have physiological mechanisms that make them seek intensive contact," de Waal added, "and that perhaps provide them with satisfaction from such contact, the question of how these mechanisms evolved still remains."[51] The powerful "sisterhood" of female bonobos, in stark contrast to the male-dominated chimpanzee realm, raises intriguing questions about the origins of their unique social world and our own.

(Opposite) Bonobos share meat at the LuiKotale reserve in the Democratic Republic of Congo. Unlike males, female bonobos regularly share food and engage in sex with one another to build alliances that may even extend beyond their immediate community.

FIRST OF OUR FAMILY

It was the first day in the field for graduate student Chalachew Seyoum, newly arrived at the Ledi-Geraru Research Project in Ethiopia's remote Afar region. Here, for nearly a decade, a fossil hunting team had returned each season for what co-leader Kaye Reed says they called "death marches"—daily treks in blistering heat that often left them empty-handed. On that morning in January 2013, Seyoum remembers that he was "full of energy and with a fresh eye"[52] as he joined the team in surveying a lunar-like landscape of desert hills. As Seyoum picked his way around the top of a hill, a shining piece of tooth enamel sticking up out of the dirt caught his eye. "My heart skipped a beat," he says, "as I realized that it was part of a hominin tooth—not only that, but it was actually intact along with a piece of jaw." Nearby, he spotted a second piece that fitted the first. He was holding the complete, well-preserved left half of a hominin jawbone. "I felt stunned but also very calm."[53]

No less stunning was the age of the jawbone, sandwiched between precisely dated layers of volcanic ash. At 2.8 million years old, it falls in the middle of a crucial gap in the fossil record that marks the emergence of *Homo,* the genus to which we belong. In that gap are a few scraps of fossils "that would fit in a shoebox and leave room for a decent pair of shoes," as paleoanthropologist Bill Kimbel remarked.[54] On one side of the gap, dating just over 2.3 million years ago, were the oldest *Homo* fossils; on the other, at around 3 million years old, were *Homo's* likely ancestors: australopiths like Lucy, discovered at the Hadar site in the Afar region less than 20 miles away. The Ledi-Geraru jaw fragment promised to bridge this vital gap in the story of our origins.

"Recalling the age of the sediments," Seyoum says, "I knew that it was going to be a very important specimen."[55] Back at camp that afternoon, the team took measurements of the jaw that fell well beyond the average of Lucy's species, *Australopithecus afarensis*. Subsequent analysis drew attention to its advanced features, for example, slimmer molar teeth and a more evenly proportioned shape than the typical *A. afarensis* jaw. Yet its primitive, sloping chin resembles that of a Lucy-like ancestor. "The Ledi jaw helps narrow the evolutionary gap between *Australopithecus* and early *Homo*," Kimbel notes. "It's an excellent case of a transitional fossil in a critical time period in human evolution."[56]

Was *Homo*'s emergence also marked by new behavior, skills, and adaptations? In 1964, Louis Leakey and Phillip Tobias had announced their famous discovery of *Homo habilis*, or "handy man," based on fossil hand bones and skull fragments from Olduvai Gorge dating back to 1.75 million years. Louis Leakey was convinced that the key traits distinguishing us from the living apes—notably, our big brain, toolmaking, and dexterous

(Opposite) Chalachew Seyoum holds the 2.8-million-year-old jaw that he discovered on the morning of his arrival at Ledi-Geraru in Ethiopia's Afar region. The fossil is a potential candidate for an ancestor in transition from *Australopithecus* to early *Homo*.

(Opposite) The Ledi-Geraru jawbone has slimmer molars and more even proportions than that of *Australopithecus*. These and other traits suggest it should be identified as early *Homo*.

hands—all emerged in a single "package" that marked a decisive turning point in our origins. The idea that many of our most essential human attributes appeared together, sharply divided from their primitive roots, has remained strong, mainly because of that yawning gap in the early fossil record.

Yet new clues are pointing to a more subtle picture. Recent evidence that stone tools appeared well before 3 million years ago (see p. 94) suggests that australopiths, not early *Homo*, may have begun crafting them. Important features of our anatomy, including upright walking, an expansion in brain size, and human-like wrists and hands, had already appeared in Lucy's species. Based on such evidence, some scientists seek to dispel the notion of a sharp transition point or a single "package" of advanced capabilities and behavior, and instead see key stages in our development as spread out across millions of years. If we search for the origins of *Homo*, argues Bill Kimbel and his colleague Brian Villmoare, we find it rooted "in ancient hominin adaptive trends, suggesting that the 'transition' from *Australopithecus* to *Homo* may not have been that much of a transition at all."[57]

While *Homo*'s major adaptations may have appeared piecemeal over a vast period of time, it need not rule out specific episodes of rapid change. Indeed, Kaye Reed, the co-leader of the discovery team, thinks that the Ledi-Geraru jaw may, indeed, signify such an episode. She points out that less than 200,000 years separate the last fossils of australopiths from the jaw, with its distinctly *Homo*-like shape and molars, which likely indicate a shift in how the Ledi-Geraru hominin procured its food. "What we want to know," she says, "is what possibly influenced this relatively fast change in diet."[58]

While a single jaw fragment is obviously limited in what it can tell us, no one doubts importance of the Ledi-Geraru find, which represents the earliest known appearance of *Homo* traits in an era almost devoid of other fossils. We owe the discovery to Chalachew Seyoum's sharp eyes and expert field skills. Those skills might never have developed without The Leakey Foundation's Baldwin Fellowship, which enabled him to pursue a doctorate at Arizona State University. The Fellowship continues to open doors for scholars like Seyoum, who faced limited opportunities in Ethiopia but was determined to pursue the extraordinary richness of his nation's fossil past.

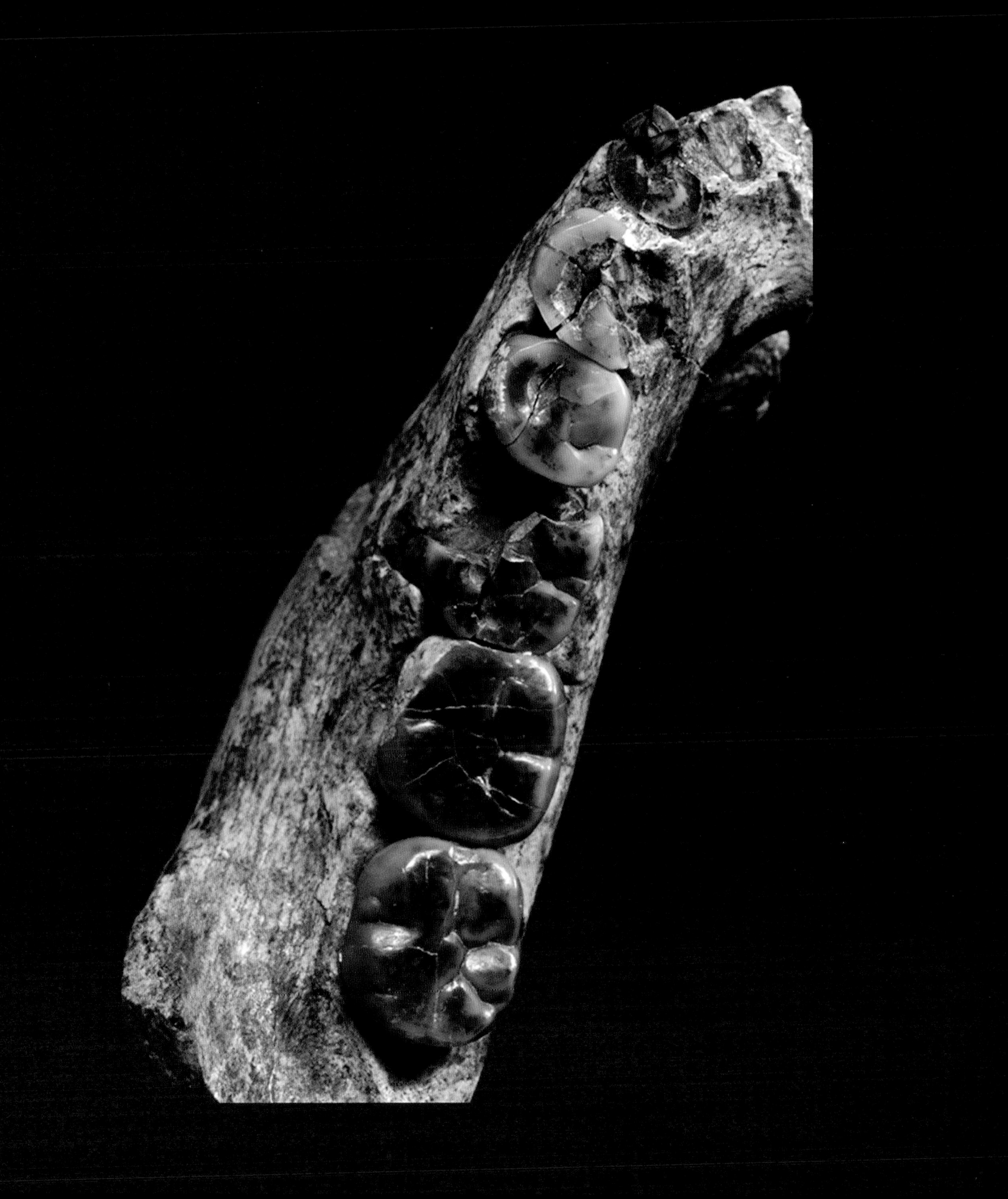

THE RISE AND FALL OF 1470

One of the most influential and fiercely debated fossil discoveries began unpromisingly. Richard Leakey recalls it as "just a few scraps of bone being eroded out of sandy sediments in a steep gully" near the shores of Lake Turkana in northern Kenya "that now looks more like a moonscape than a place that once supported a vital stage in our ancestral history."[59] In July 1972, Bernard Ngeneo, a young member of the Leakey team, bent down to examine what his colleagues had cursorily dismissed as antelope bones. Ngeneo recognized them as hominin.

Back at the research camp at Koobi Fora, it took Meave Leakey six arduous weeks to piece together the 150 fragments, at one point exclaiming in frustration that "we have a jigsaw puzzle with no edge pieces." What finally emerged was an exceptionally complete cranium, designated KNM-ER (Kenya National Museum-East Rudolf) 1470. "It was larger than any of the early fossil hominins I had seen," wrote Richard, "but the question was, how large was the brain?"[60] Eagerly, he and his colleague Bernard Wood plugged the gaps in the skull with tape and modeling clay, and filled it with sand to measure its volume. It proved to be around 800 cc, confirming that 1470 had a much bigger brain than that of any other fossil of its time, more than 2.6 million years ago. "Historic moment!" Wood wrote in excitement in the camp journal.[61]

(Opposite) Meave Leakey pieced together the KNM-ER 1470 skull from 150 fragments, here seen surrounded by additional fragments that she could not fit into the reconstruction. Its discovery played a pivotal role in debates during the 1970s about the origins of *Homo*.

During the late 1970s, popularized by Richard Leakey in a best-selling book and television series, the 1470 skull became *the* iconic face of our ancestors in the media, rivaling the popularity of Lucy. Not only its big brain but its long, wide, flat face was completely different from the smaller, more protruding, ape-like faces of other fossil humans from East and South Africa. While 1470 once had large teeth, it lacked the massive jaw and huge chewing muscle attachments of the "nutcracker" *Paranthropus* (see p. 91). All these differences led Richard to some electrifying conclusions. Evidently, he wrote, "the human ancestral line, *Homo*, must have emerged much earlier than most people suspected, earlier perhaps by as much as a million years."[62] Since it co-existed with the early australopithecines, it was unlikely that those more ape-like creatures could be our direct ancestors—"cousins, yes, but descendants, no."[63] A little over two weeks after Wood's "historic moment," Richard flew down from Koobi Fora to Nairobi with the groundbreaking new skull and showed it proudly to his ailing father, Louis. It was a triumphant vindication of Louis's lifelong quest to find a big-brained, tool-using human ancestor in East Africa's fossil past. Just five days later, Louis died of a heart attack in London.

Although convinced that 1470 was our direct ancestor, Richard Leakey was cautious about assigning it to a specific species. The most obvious candidate was *Homo habilis* or "handy man," the tool-using ancestor that his parents had discovered at Olduvai Gorge back in the early 1960s. Although it was nearly a million years older and had a bigger brain than the Olduvai specimens, many of Richard's colleagues identified 1470 as *Homo habilis* and accepted that it was probably the earliest *Homo.*

Then, in the late 1970s, the story grew more complicated. A re-examination of the volcanic layer in which 1470 was found led to a new date for the fossil of only around 1.8 million years. Abruptly, 1470 was dethroned from its title of earliest ancestor. The younger age made it not only contemporary with *Homo habilis* finds at Olduvai and elsewhere but with a more advanced species—the forerunners of *Homo erectus* in Africa and Asia. 1470, then, was simply too late to be a likely direct ancestor of our lineage.

Moreover, the identity of *Homo habilis* itself began to unravel. Originally, it was assumed to be a single species that likely gave rise to *Homo erectus* and our own lineage, yet most of the relevant fossils were scrappy and too few and far between to be sure. As new discoveries slowly accumulated, the *habilis* label grew increasingly contentious; to many, it seemed less like a single group than a hodge-podge of different creatures all shoehorned together. In 1986, Russian anthropologist Valerii Alexeev proposed that 1470 should be broken out into its own separate species called *Homo rudolfensis*, after the colonial name for Lake Turkana. When Bernard Wood later re-examined 1470 and the other fossils, he, too, supported the proposal. The East African fossil landscape at the dawn of humanity some 2 million years ago was now beginning to look crowded. In place of a single hominin species, there had been no less than four, all co-existing together: *Homo rudolfensis*, *Homo habilis*, *Homo erectus*, and the "nutcracker" *Paranthropus boisei*.

The quest for a single big-brained ancestor begun by Louis Leakey and pursued by his family over two generations has led, a half-century later, to a far more complex picture. From a single branch, we have gone to a tangled thicket of ancestors. Commenting recently on this big picture, Bernard Wood writes, "A simple, linear model explaining this stage of human evolution is looking less and less likely. Our ancestors probably evolved in Africa, but the birthplace of our genus could be far from the Great Rift Valley, where most of the fossil evidence has been found. The Leakeys' iconic discoveries...should remind us of how much we don't know, rather than how much we do."[64]

THE SKULLS IN THE CELLAR

For centuries, Dmanisi was a major crossroads and trade center along the Silk Road connecting Asia, Europe, and the Near East. It sits on a high, narrow clifftop in the Caucasus, wedged between two rivers about fifty miles from Tblisi, the capital of Georgia. An ancient cathedral and tumbled stone walls are all that remain of this prosperous medieval city; sacked by invaders, it lay neglected for six centuries. Then, in the 1980s, Georgian archaeologists began investigating the cellars of the medieval ruins. When they dug out an old grain storage pit in 1983, they made a disconcerting discovery: the tooth of a long-extinct species of rhinoceros. It turned out that the once-thriving trading post had been built on top of sediments and volcanic ash laid down 1.8 million years ago. They kept digging, and the following year they found stone tools. In 1991, they found a human jawbone—the beginning of an unexpected twist in the pivotal debate about when, and how, the first humans left Africa.

Back then, Dmanisi was also a crossroads, but of a dramatically different sort. Saber-tooth cats, European jaguars, giant cheetahs, lion-sized hyenas, and other formidable predators roamed the cliffs. Among the prey they dragged into their dens were the remnants of slightly-built human ancestors, who averaged around five feet in height and a weight of 100 lbs. While their bones bear the gnaw marks of big cats, they, too, were hunters and scavengers. With rudimentary flaked stone tools, they cut off the flesh of mammoth, deer, and wolves and ate it raw. No trace of fire has been found despite frigid winters, and they had no weapons apart from cobblestones. So what were these vulnerable little humans doing here, over 1,000 miles from Africa? The clues lay in the bones. Over the next 15 years after the jawbone find, with major support from The Leakey Foundation, a team led by David Lordkipanidze recovered skeletal parts and no less than five well-preserved skulls. One of them, Skull 5, is among the most complete early human adult skulls in the fossil record.

Although dug from the same geologic layer, they represent a surprising range of shapes and sizes, and a tantalizing mix of primitive and advanced features. Skull 5 has a massive, rugged, protruding face, in marked contrast to the rest, notably the much smaller and slighter Skull 3. It had survived long enough for its big teeth to be worn down to stumps and its jaw to be inflamed with arthritis. Despite their variety, all five had relatively tiny brains; the braincase of Skull 5 is only a third that of a modern human's. Two partial skeletons share legbone proportions that are essentially modern, yet the shoulder blades are a more ape-like shape, probably inherited from tree-climbing ancestors. The debate centers on whether all five skulls belong to several species,

(Opposite) One of the most astonishing discoveries of the 1990s was of slightly built, small-brained hominins at Dmanisi, Georgia. Lacking fire and equipped with rudimentary tools, they left Africa and reached the Caucasus 1.8 million years ago.

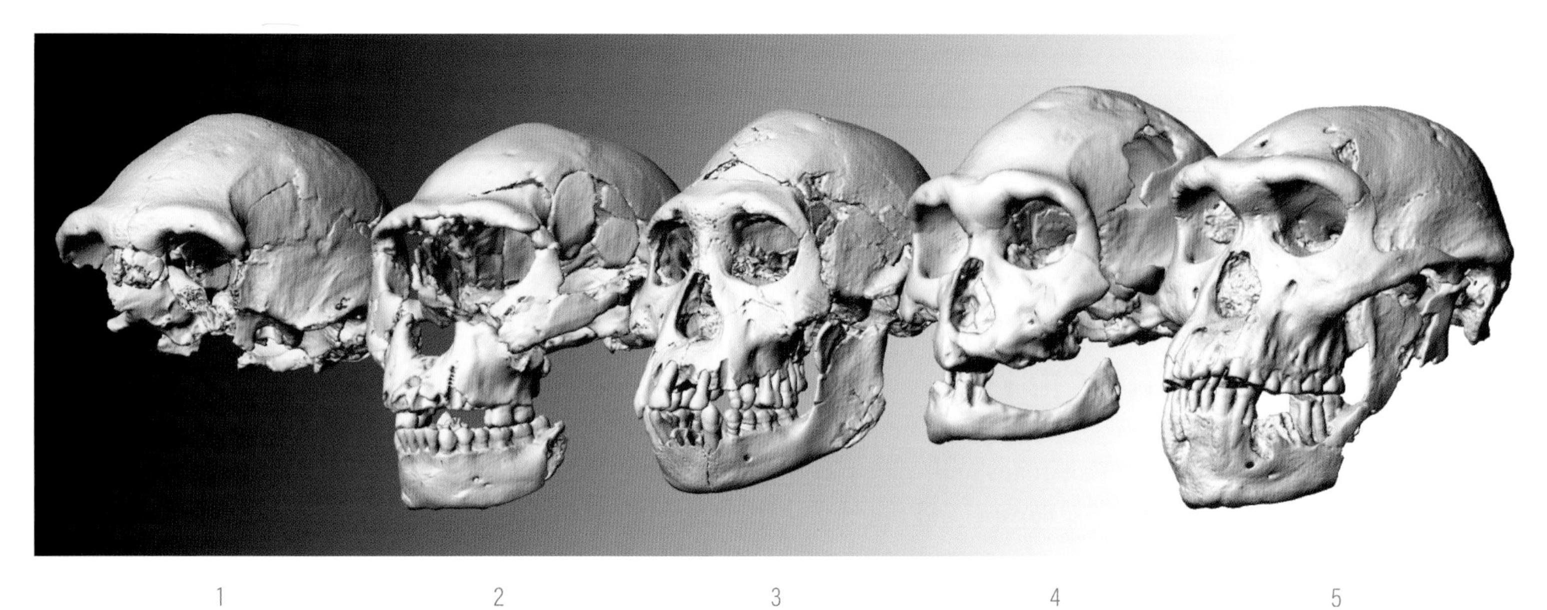

1 2 3 4 5

Five well-preserved skulls were found at Dmanisi (computer reconstruction above). The slightly built Skull 3 is in striking contrast to Skull 4, an elderly, toothless male, while the massive Skull 5 has a brain little bigger than a large chimpanzee's. (Opposite) An air view of Dmanisi shows the medieval ruins in the foreground.

or just one—and whether that one is related to *Homo erectus*, *Homo habilis*, or represents something in-between. While their small stature and brains resemble the more primitive *H. habilis,* their legs and some skull and dental features are mostly a better fit for the more advanced *H. erectus*.

However they are classified, the essential point is that the Dmanisi hominins are dated *before* the classic specimen of *Homo erectus*, the 1.6-million-year-old "Turkana Boy" discovered in East Africa in 1984 (see p. 122). Prior to the Dmanisi finds, it was assumed that our ancestors were incapable of leaving their African homeland without some of the innovations often credited to *Homo erectus* such as advanced hand axe technology and the mastery of fire. Yet the absence of big brains, fire, and capable tools was evidently no obstacle for their survival in the Caucasus, where they faced more extreme seasonal temperatures and far less abundant plant foods than they would have in Africa. "This was not an easy environmental transition to make," notes paleoanthropologist Ian Tattersall, "and it almost certainly could not have been managed by a typical primate species. Evidently the key to early hominins' success in Eurasia, then as now, was the unusual flexibility of behavior that has always been the hallmark of their African ancestors."[65]

Poignant evidence of such unusual behavior is apparent in Skull 4, that of an adult male who lost all but one of his teeth well before he died. Most of the empty tooth sockets in his jaw had shriveled away, a process that would have taken several years. Chewing raw meat would have been an ordeal, yet the fact that he survived means that other members of the community probably helped prepare food for him. If so, this would be by far the oldest known example of compassionate care of a disadvantaged individual known in the annals of prehistory.

THE CASE OF THE SHRINKING ANCESTOR

Kamoya Kimeu, head of the National Museums of Kenya's "hominid gang" of fossil hunters, was feeling restless. While the rest of the team slept through a hot afternoon in August 1984, Kimeu headed for a stony slope beside the Nariokotome River, near Lake Turkana. "I saw that area and I said, 'I must look there.'"[66] Then his eagle eye caught a matchbox-sized piece of bone that he recognized as a fragment of hominin skull. When the team began screening the soil surrounding the spot, excitement grew. "We couldn't believe it," Meave Leakey told NOVA, "but we started getting pieces of ribs. These were the parts of *Homo erectus* that nobody actually knew about, nobody had ever seen before, so every bone that came out of the ground was something brand new to science."[67] Over several seasons, Leakey's team recovered some 150 pieces comprising the most complete fossil of *Homo erectus* ever discovered. What emerged from the lab was startling: the assembled skeleton indicated that a dramatic step toward our modern human body form had taken place some 1.6 million years ago. "Turkana Boy" also raised intriguing questions about the emergence of humanity's unusually long period of childhood and adolescence.

The story of *Homo erectus* begins in 1891–92 with a landmark discovery in Java, Indonesia. Eugene Dubois, a Dutch physician and anthropologist, set a team of convict laborers to work at a site near the Solo river. They unearthed a thick skullcap, a few teeth, and a human-like thigh bone. While the low-roofed skull and its big brow bridges struck Dubois as primitive, the modern-looking thigh bone convinced him that the "ape-man" was an upright walker, so he classified the discovery as *Pithecanthropus erectus*, or "upright ape-man." Subsequent finds, notably the celebrated Peking Man fossils unearthed in China during the 1920s, established *Homo erectus* as the first human ancestor to populate Asia. Before the Leakeys made their pivotal discoveries at Olduvai Gorge, it looked as if Asia, not Africa, might be the birthplace of humanity. Eventually, fossils with similarities to the Asian *erectus* were found in Africa. (To distinguish them from their Asian counterparts, some experts refer to the African *erectus* as *Homo ergaster* or "work man," after the distinctive stone tools known as hand axes found at many of their sites.) By the time of Turkana Boy's discovery in 1984, *Homo erectus* was seen as an innovative toolmaker and trailblazing colonizer, the first ancestor to leave Africa around 1.9 million years ago, venturing rapidly across southern Asia and eventually reaching Indonesia.

Turkana Boy's impressive skeleton established the iconic image of *Homo erectus.* Although the skull has a flatter, less projecting face than a typical ape's, it housed a brain of around 880 cc, not much bigger than half the average human size today.

(Opposite) The classic *Homo erectus* fossil known as Turkana Boy, here seen in a reconstruction, marks a striking advance toward an essentially modern body plan.

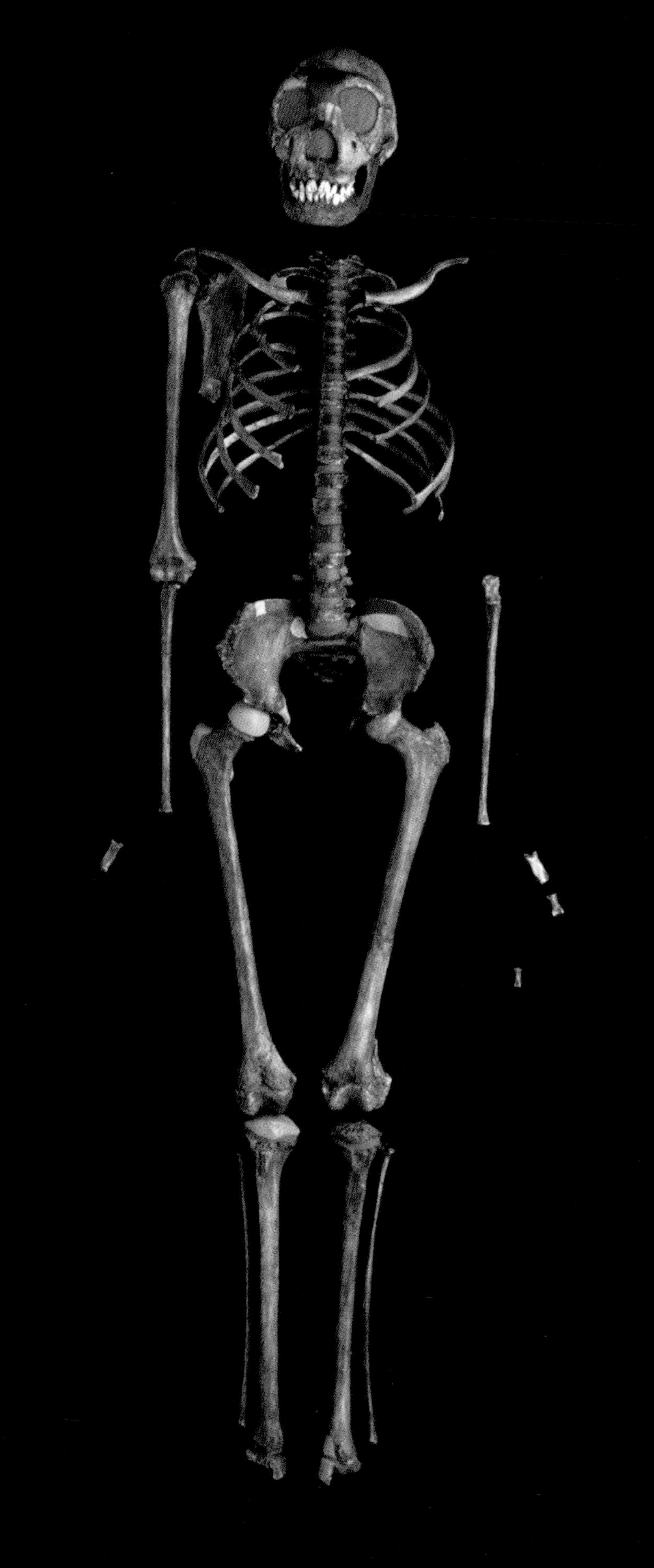

(Opposite) Turkana Boy's phenomenally complete skeleton was assembled from around 150 fragments, including delicate, rarely preserved rib bones. Despite his many modern features, the Boy had a brain only half our size.

While twice the size of a chimpanzee's or an australopithecine's, this capacity is already reached by human infants by the end of their first year. Turkana Boy's mental world may well have been very different from our own.

Yet, from the neck down, his skeleton is close to the modern pattern and in striking contrast to his stocky, pot-bellied predecessors, the australopithecines and *Homo habilis.* Instead, his long and lanky proportions, with no lingering hints of tree climbing abilities, point to a body adapted for a hot, dry climate. His robust leg bones testify to a committed, striding biped, capable of walking and running long distances. A set of footprints from around the same period left in volcanic ash at Ileret, Kenya, are presumed to be those of *Homo erectus*. They indicate a walker similar in height and step to an average North American today.

In fact, the height of Turkana Boy was one of his most arresting features—or so it appeared at first. Measurements of his long bones suggested that he would have been about 5 foot 1 inch at the time he died. But how old was he? Most of his teeth had not yet completely formed; X-rays showed that their roots were still growing. If we match these clues with the dental development of today's children, we find that he would have been 11–12 years old and about to enter his adolescent growth spurt. Had he survived, he might have reached an adult height of no less than 6 foot 1 inch—a giant compared to *Homo habilis* or *Homo rudolfensis,* some of whom were probably a scant 3 feet tall.

But what if Turkana Boy grew at a pace like that of chimpanzees, much faster than today's children? In that case, he would have been younger at death and shorter as an adult than previous estimates. To investigate this possibility, Turkana Boy's teeth were re-examined with a more precise technique. Tooth enamel is laid down in a regular daily pattern that under a microscope can be counted like tree rings, then compared with the record of living children and chimpanzees. When assessed along with other anatomical clues to the boy's age at death, the results indicated that he was a mere eight years old. Yet, by that time, he had completed much of his adult growth. Had he survived to adulthood, he would probably have reached between 5 foot 4 inches to 5 foot 10 inches. While much uncertainty and debate hedges these calculations, there is little doubt that Turkana Boy was following a unique trajectory, somewhere in between the accelerated ape pattern and our unusually lengthy childhood today.

The earlier evidence for Turkana Boy as a strapping, six-foot ancestor with an advanced, athletic-looking build had helped fuel a popular image of *Homo erectus* as a trailblazer, well prepared to conquer lands far beyond the African continent. Since the revisions to his estimated height and age, evidence has accumulated that the first African exodus may have begun far earlier—perhaps even half a million years earlier—than Turkana Boy. The real pioneers may have been a less developed form of *Homo erectus* with more primitive bodies and simpler tools, similar to the hominins at Dmanisi. In that case, it could be that crucial stages in the early evolution of *erectus* happened somewhere in Asia rather than Africa, then later spread back to the Rift Valley. While this remains unproven, it would be ironic if Turkana Boy assumes a new popular image as immigrant rather than trailblazer.

While all these developments have—literally—taken Turkana Boy down a few notches, nevertheless, the significant shifts that we see in his anatomy signal an ancestor that had taken giant strides on the path toward modern humanity.

BORN, AND EVOLVED, TO RUN

It began with a pig on a treadmill. In 1991, Dennis Bramble, an expert in animal locomotion, was visiting a Harvard lab where Daniel Lieberman, then a Ph.D. student, was coaxing a pig to trot. As Bramble watched the experiment, he exclaimed, "You know, that pig can't hold its head still!" Lieberman says, "This was my 'eureka!' moment. I'd observed pigs on treadmills for hundreds of hours and had never thought about this."[68] Bramble had a theory for why a pig's head bobs around at the trot: unlike proficient runners such as horses, dogs, and humans, pigs lack a springy ligament that connects their skulls to their spine, steadying their heads. Bramble showed Lieberman where the ligament attaches to the base of a human skull; at that spot, there's a slight but sharp ridge—a bony signature of our ability to run. That raised a couple of big questions: How far back did that signature go in the fossil record? And just how important was running to the evolution of humans?

The questions were of special interest to Lieberman, an avid jogger since his teens and now a Harvard anthropologist. He and Bramble developed an influential theory that sees the development of running as a pivotal point in human evolution. Regardless of our personal athletic prowess or lack of it, we all share a unique set of adaptations that makes us highly efficient long-distance runners. "We're loaded top to bottom with all

(Opposite) Arnulfo Quimare is a celebrated ultramarathon runner of the Tarahumara people of Chihuahua, Mexico, known for their prodigious feats of long-distance running wearing tire tread sandals.

(Opposite) Until recently, the San people of the Kalahari desert still relied on traditional hunting skills such as endurance running. Studies show their running performance compares with that of the world's best ultramarathon runners.

these features," Lieberman notes, "many of which don't have any role in walking."[69] They range from the long, elastic tendons in our legs and the springy arches in our feet to our big knee joints and hefty gluteus maximus, which is crucial for balancing the body during a run: with each forward stride, it contracts, pulling the torso back and preventing us from falling.

Lieberman argues that humans are, in fact, highly tuned running machines to a degree that puts most other species to shame. While horses, dogs, antelope, and hyenas all gallop much faster than we, only humans can run marathons. Most African animals cannot run at all in midday heat. But humans, with our unique hairless bodies and abundant sweat glands, can keep up a steady jog for hours.

Long-distance running, argue Bramble and Lieberman, enabled our ancestors to leave the safety of the trees and successfully hunt and scavenge in the open savanna. Before the invention of projectile technology, one key method to bringing down an antelope was to track and pursue it across the landscape until it collapsed from overheating. A modest human running speed is fast enough to drive an antelope from a sustainable trot into a gallop, and at that pace, it can no longer lose heat by panting. "Even I, a middle-aged Harvard professor," Lieberman says, "have run down jackrabbits and horses over long distances. One way is to make them gallop. As soon as an animal gallops, it cannot pant."[70]

Requiring skilled tracking and cooperation, this technique of endurance hunting is well-documented among recent hunter-gatherers, notably the Tarahumara of northern Mexico and the San of the Kalahari Desert. Groups of San hunters have been observed pursuing antelope in 100° F temperatures, keeping up a steady pace similar to marathon runners for several hours. They do not need weapons to bring down prey, since heat exhaustion alone will finish the animals off.

Two million years ago, our defenseless ancestors on the open savanna had no projectile weapons, yet they had to compete for meat with fearsome lions, cheetahs, and hyenas. Around that time, the appearance of *Homo erectus*—notably the Turkana Boy with his long, lanky, and athletic build—marks the arrival of long-distance running adaptations for the first time in the fossil record, including that telltale bump on the skull that helps steady the head. It was a vital step along the long road of humanity's emergence. Only endurance running and persistence hunting, Bramble and Lieberman argue, can explain how our ancestors gained regular access to meat sources that were crucial to fueling the expansion of our big brains.

THE GREAT ANCESTRAL BAKE-OFF

A dinner date with wild chimpanzees is something to avoid. Quite apart from their lack of table manners, the menu poses significant challenges. The main entrée is wild forest fruits, typically "very unpleasant," notes Harvard biological anthropologist Richard Wrangham, renowned for his chimp studies. "Fibrous, quite bitter. Not a tremendous amount of sugar. Some make your stomach heave."[71] In Kenya's Gombe National Park, where Wrangham began his career as a graduate student under Jane Goodall, chimps spend more than six hours a day chewing fruit, leaves, and bark. Even if we wanted to, our relatively puny guts could not process all that roughage, nor could we survive on it. In fact, eating only uncooked food can present problems for today's raw food enthusiasts. In one study, more than 50 percent of women restricted to raw foods failed to ovulate, which is clearly an issue from an evolutionary perspective. Cooking is evidently essential for human survival, but how and when did we make the great leap between primate and human cuisine? How did Chez *Pan* become Chez Panisse?

One evening in 1997, Wrangham was pondering this question beside the fire in his Massachusetts home when he began thinking about all the advantages that cooking could have brought to our ancient ancestors. Could it be *the* crucial innovation that set our human ancestors apart from the apes and launched them down their unique evolutionary pathway? To find out, Wrangham teamed up with biologists on innovative lab experiments that showed just how well-adapted humans are to eating cooked food and the huge advantages cooking brings. It softens and breaks down foods, enabling us to spend less than an hour a day chewing compared to a chimpanzee's six. It also kills toxins in foods, and makes nutrients more available to the enzymes in our stomach. Wrangham's studies indicate that, on average, cooking meat or plants results in an overall energy gain of around 30 percent. This leap in calorie intake was vital for our evolutionary success, he believes; it "reduced feeding time, freeing men to hunt, lowered weaning time, creating bigger families; allowed brain size to increase...It was so important, that it likely drove the evolution of our genus *Homo*. Basically, if the cooking hypothesis is right, it turned us from advanced ape to early human."[72]

When did this breakthrough happen? Wrangham argues that the advent of cooking was essential for fueling the enlarged brain

(Opposite) A group of nomadic Hadza hunter-gatherers in northern Tanzania cook a bird over a campfire.

and body of *Homo erectus,* which first appeared in Africa around 1.6 million years ago. The change is most obvious in the spectacular Turkana Boy fossil from East Africa (see p. 122), which is taller and more modern-looking than any prior ancestor. He also had much smaller jaws, teeth, and guts, which suggest there was no longer a need for endless chewing. Was *Homo erectus* cooking his supper?

Plausible as the theory is, proving it is another matter. For archaeologists, distinguishing primitive cooking hearths from the debris left by natural wildfires can be tricky. Intriguing evidence of deliberate fire control was recently discovered at a site at Koobi Fora, Kenya, close to where Turkana Boy was found, and dating to the same period. High-resolution mapping and soil analysis revealed burned bone fragments, heat-cracked rock, and reddened soil in patterns consistent with human fire activity. At Wonderwerk Cave in South Africa, similar evidence of burned bone and rock together with plant ashes was uncovered far back from the cave entrance suggesting that ancient humans deliberately built fires there about a million years ago.

Even though a handful of early sites hint at fire-related activities that could have involved cooking, the broader picture poses problems for the theory. When ancestral humans began leaving Africa, they faced more extreme climate and energy demands as they moved into the higher latitude regions of Europe and Asia. Yet, none of the earliest European occupation sites such as Dmanisi, in the Caucasus, or Atapuerca, in northern Spain, have produced significant quantities of burned bone or other signs of hearths.

Clear evidence for the regular use of fire only appears in Europe and the Near East much later, some 400,000 years ago, and even then the picture is puzzlingly inconsistent. At two Neanderthal sites in southwest France, investigated with support from The Leakey Foundation, thick charcoal and burned bone layers were followed by a long period when there is no evidence for hearths—even during some of the most frigid episodes of the last Ice Age, when the Neanderthal hunters tracked reindeer across a bleak, tundra-like landscape. Could they have survived on raw meat, despite Wrangham's influential research indicating that it would be next to impossible? Some critics of his theory suggest that before regular cooking, our ancestors might have got by on uncooked energy-rich animal parts, such as internal organs, bone marrow, brains, and intestines—for our tastes, a diet even less appetizing than a night out with wild chimpanzees.

(Opposite) A Hadza mother prepare a meal of raw meat for cooking. After the invention of stone tools, cooking was perhaps humanity's next most crucial innovation, vital for extracting energy efficiently from foods and fueling our big brains.

MOUNTAIN GORILLA MYSTERY

Ever since the 1933 film *King Kong*, popular culture has depicted the gorilla as a ferocious, lustful monster, the ultimate symbol of unchained male aggression. With their heavily muscled, 350-pound bodies and dagger-like slashing canines, male mountain gorillas, or silverbacks, are certainly formidable creatures. Yet in the 1970s, Dian Fossey's foundational work at the Karisoke Research Center, in central Africa's Virunga Massif, helped establish a totally different image of gorillas as "gentle giants," largely peaceable vegetarians that spend most of their days browsing and resting in the cloud forests. In the groups that Fossey typically observed, a single dominant silverback held court over a harem of a dozen or so females and juveniles. Since silverbacks have to fight for exclusive access to multiple females and to protect the group from outsiders, their huge fangs and outsize bodies are important assets. But these fights mostly involve noisy bluff and bluster and only rarely inflict serious injury.

At least, that was the rule until the 1990s, when the social order of some of the Virunga gorillas changed dramatically. In some groups, it became routine to see half-a-dozen or more males co-existing together in relative harmony. In one case, observers counted 65 individuals in a single group—an unthinkable situation in Fossey's era. A new generation of gorilla researchers is determined to figure out what is going on. In 2003, Stacy Rosenbaum was invited to assist with a major study of male gorillas supported by The Leakey Foundation. She trekked for thousands of hours in Karisoke's forest, recording their behavior and collecting fecal and urine samples for a huge "biobank" of hormone and DNA data. The goal was to decipher the social rules that govern the large multi-male groups.

As the study progressed, Rosenbaum became increasingly fascinated by the relationships between males and infants. When they are old enough to be independent of their mothers, playful youngsters are drawn "like magnets," she says,[73] to the company of dominant males, who indulge them in rough-and-tumble games, hold and groom them affectionately, and allow them to sleep in their nests. "Here are these huge animals that are clearly strongly adapted for fighting," Rosenbaum adds. "In spite of that, they rarely fight and can be extraordinarily gentle and nurturing. It's amazing to see them pick up infants and be incredibly loving—for want of a better word."[74] Their tender behavior with infants is in striking contrast to many other primate societies with ranked dominant males, including those of baboons and chimpanzees, who have scant interest or involvement with infants.

Could their behavior simply be due to paternal instincts? With the help of The Leakey Foundation, Rosenbaum returned in 2011, by then a UCLA graduate student, to answer this

(Opposite) Belying their aggressive stereotype, dominant male gorillas at the Karisoke Research Center seem to dote on young infants that seek them out for playtime and grooming.

question. Gorillas are not monogamous, and in a small-group harem with a single male, nearly all the infants will be sired by the resident silverback. But in a bigger, multimale group, some of the young will have other fathers. That raises the question of whether paternity matters or is even recognized in the first place. Rosenbaum's observations show that, for their playdates, the youngsters regularly seek out the highest-ranked males, who may or may not be their biological fathers.

The actions of the young gorillas make sense because dominant silverbacks provide the best protection for the highly vulnerable infants, over a quarter of whom die by age three. Many fall victim to predators, which are mostly other gorillas. When a solitary male outsider tries to insinuate himself into the group, he will often kill infants so that nursing mothers will stop lactating and become available for mating. Infanticide is the dark side to the gorilla's "gentle giant" image.

As for the silverbacks, they either do not recognize their own infants or, when it comes to interacting with them, paternity is not an issue. "My suspicion is that they *do* know perfectly well that they're not the father," she says, "but they don't care. They enjoy playing with infants; it's something that they like to do and get pleasure out of."[75] As well as the joy of playtime, hidden instincts could also be at work. In a recent preliminary study, Rosenbaum finds that the males who are most attentive to the young are also the most reproductively successful. This could mean that females consider a doting father to be a more attractive mate—a possibility we might recognize from our own experience.

Overall, Rosenbaum's evidence points to a surprising degree of male involvement with the younger generation, greater than would normally be expected in a species so strongly focused on dominant males. The social flexibility of these highly intelligent creatures, so far from the archetype of the monster ape, has implications for understanding our own world as well as our past. In a recent essay, Rosenbaum asked, "If the gorilla can transform its entire social structure, flipping highly gendered behaviors totally around—and flourish in the process—can we not also?"[76]

(Opposite) At the Dian Fossey Gorilla Fund's Karisoke Research Center, Stacy Rosenbaum and Gorilla Program Manager Jean Paul Hirwa extract hormones from feces. The hormone data helps deepen understanding of how social relationships between mountain gorilla infants and adult males may have evolved.

GORILLA BONE DETECTIVES

Dian Fossey's friends shouldered her plain white pine coffin up the steep, muddy trails to the Karisoke Research Center, located in central Africa's Virunga Mountains, to a small field studded with simple wooden grave markers. Until her brutal murder five days earlier, on December 26, 1985, Fossey had devoted 18 years to her celebrated study of wild mountain gorillas. Now, she would lie alongside over a dozen of the gorillas that she had known so intimately. She was buried only a few steps away from Digit, the young silverback she had loved the most.

Apart from honoring their lives, Fossey had buried these gorillas to preserve their bones for future study. Over nearly a decade, she had also shipped a total of 15 mountain gorilla skeletons and ten skulls to the Smithsonian's National Museum of Natural History. After her death, trackers from the Dian Fossey Gorilla Fund and veterinarians from Gorilla Doctors and the Rwanda Development Board buried the gorillas wherever they found the bodies in the forest. Eventually, they created a new cemetery next to national parks headquarters, where a careful protocol was worked out to ensure that each skeleton would be preserved and readily identified when later dug up for study.

During the 2000s, Antoine Mudakikwa, Senior Veterinarian with the Rwanda Development Board, grew interested in the idea of starting up such a study. A meeting with U.S. biological anthropologists including Shannon McFarlin helped get the project off the ground, as did the help of The Leakey Foundation and other partners. A striking difference between great apes and humans is our long, slow period of early childhood development, followed by a growth spurt in adolescence. To help understand how this unusual pattern evolved in our ancestors, the team wanted to investigate the equally poorly understood process of gorilla growth and maturation and how it is affected by changes of diet and social behavior. They hoped to discover clues to development in the gorilla bones that might also apply to the fossil bones of human ancestors.

In 2008, the project's team—Mudakikwa, McFarlin, other biological and forensic scientists, and their colleagues from the Fossey Fund, Gorilla Doctors, and Rwanda Development Board—began trekking through the mountains to locate the unmarked forest graves. During their first season alone, guided mainly by the recollections of rangers and veterinarians, the team managed to locate over 70 skeletons. In later seasons, they exhumed bones from the current cemetery, steadily adding to a collection that now stands at over 150 individuals. The identities of more than half of them are known and can be linked to the meticulously detailed life histories and records of behavior and health maintained by the Fossey Fund, Gorilla Doctors, and the Rwandan National Park Service for over 50 years. The collection offers scientists an unprecedented combination of forensic information and gorilla "biographies."

(Opposite) In the cemetery at the Karisoke Mountain Research Center, Dian Fossey's grave lies next to her beloved silverback gorilla named Digit.

DIGIT
1965_1977
"NYIRAMACHABELLI"
DIAN FOSSEY
1932 – 1985
NO ONE LOVED GORILLAS MORE
REST IN PEACE, DEAR FRIEND
ETERNALLY PROTECTED
IN THIS SACRED GROUND
FOR YOU ARE HOME
WHERE
Dian Fossey

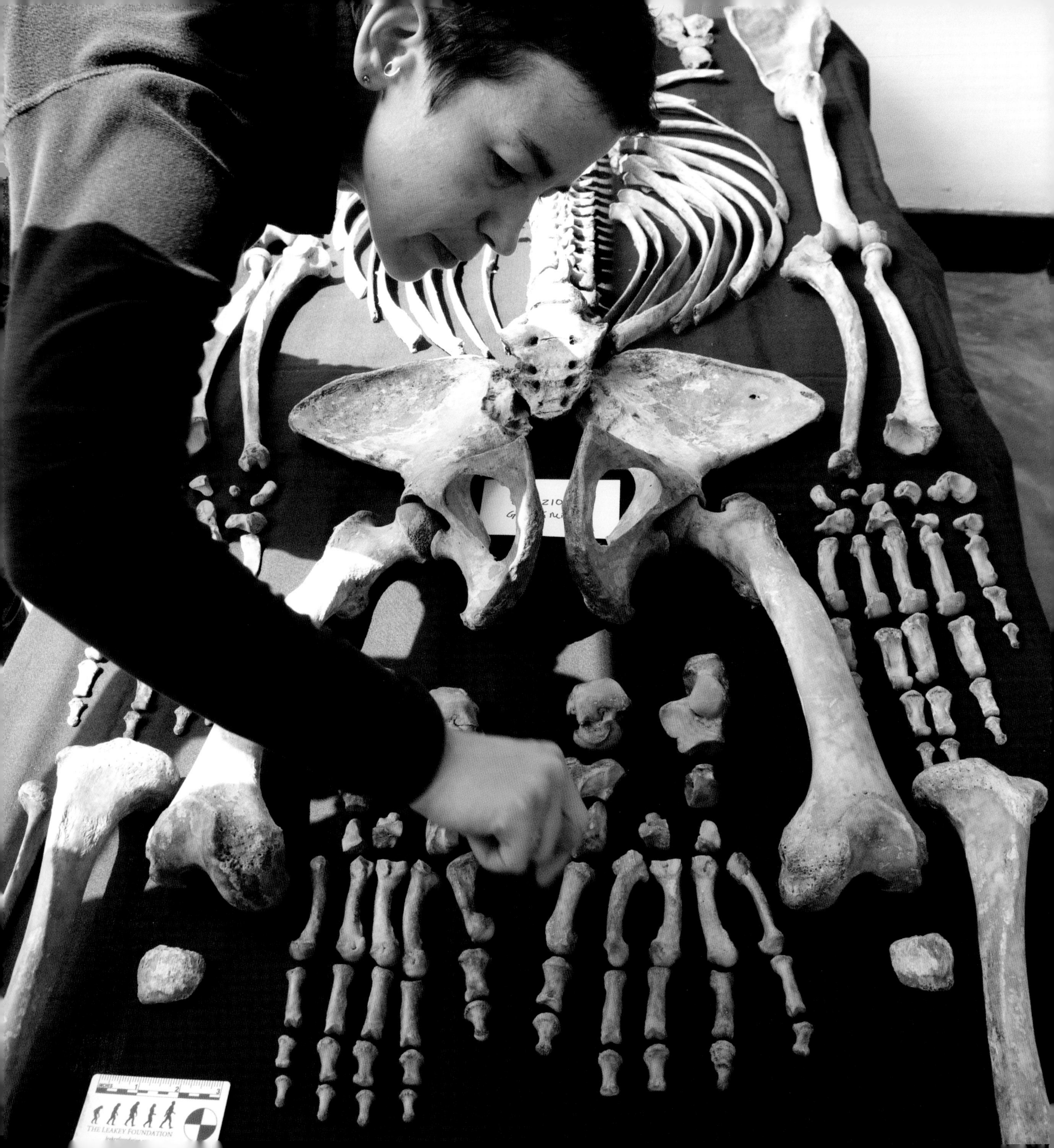
THE LEAKEY FOUNDATION

(Opposite) For over a decade, primatologist Shannon McFarlin (pictured here), Antoine Mudakikwa and their colleagues have carried out forensic analysis of the gorilla skeletons conserved at Karisoke. Their work is leading to new insights on gorilla health and growth rates, vital for the continuing preservation of this highly endangered species.

Simply judging by appearances alone can be misleading, as the team found in the case of two infant male gorilla skeletons in Rwanda's gorilla bone collection. They knew from observational records that the two infants had been born into the same group within two weeks of each other. One of them had thrived, while the other was sickly and more dependent on his mother. Both had died at age two; the healthy infant was the victim of infanticide by a male gorilla, and the weaker one died after a period of maternal separation. "When we excavated them from the cemetery," McFarlin says, "we realized that if we'd found them as fossils, we would have estimated they were six months apart in age, if not more. Their overall size and stages of tooth development were really different."[77]

To help avoid such errors, the team has used more precise ways to recognize the signature of key life stages in gorilla skeletons. Under high magnification, McFarlin and her colleagues look at cross-sections of bones and teeth to examine the microstructure of the tissues that are laid down periodically in layers, like tree rings; for example, they can detect tissue patterns typical of the fast growth spurt of adolescence. From such clues, they can estimate the rate and timing of growth in body size and tooth development, as well as recognize marks of stress, injury, and disease. This technique offers a novel way to unlock the mystery of shifting ancestral life cycles. The records of gorilla watching over five decades provide a vital cross-check that makes the skeletal technique reliable.

One drawback of those records is that they contain no measurements of the growing gorillas. That is hardly surprising; "You can't put a tape measure on a mountain gorilla," notes McFarlin.[78] So along with a colleague, Jordi Galbany, the researchers adapted a clever hands-off technique to capture the dimensions of gorillas in the forest. The principle (known as photogrammetry) is straightforward: mount twin laser pointers on a camera, point the camera at a gorilla, and two green spots will appear on the animal at a known distance apart. Take multiple photos of the same body area and a computer will calculate an accurate measurement from the data. In 2016, their first study using this technique found the intriguing result that Rwanda's gorillas grow fast compared to other wild gorilla populations at lower altitudes.

It is clear that scientists have only just begun to tap the full potential of Rwanda's unique gorilla bone collection. "The depth of information associated with these skeletons is extraordinary," Shannon McFarlin says. "We can read the bones and teeth as a record of the animal's life. They are essentially a catalog of events that the gorilla experienced, from birth to death."[79]

SECRETS OF THE ORANGUTAN CYCLE

Studying orangutans is a challenging business. As these solitary apes travel with ease through the canopy of Indonesian forests, human observers struggle to keep up, crashing through spiny undergrowth and wading through waist-high swamps. For primatologist and Leakey grantee Erin Vogel, the daily routine starts at 3 AM with a trek to the orangutan nest. "We like to get there before daylight breaks so we don't miss them coming out of their nest. Eventually they wake up and the first thing they do is urinate. It's messy and you do get urine on you and you smell bad for the rest of the day. But you smell bad anyway, because it's 105° and 98 percent relative humidity."[80] Strangely enough, the daily urine shower is more than an occupational hazard for researchers like Vogel; it turns out to be the key to unlocking an evolutionary puzzle with implications for humans as well as orangutans.

At first glance, orangutans might seem an unlikely choice for insights into our distant past. Splitting off from the great ape family tree at least 12 million years ago, they are the least closely related to us of all large living primates. Yet one striking fact makes orangutans—and their urine—an immediate target of interest to science. Orangutan mothers give birth only every six to nine years, the longest birth interval of any mammal. Seeking to understand this unusual pattern, primatologist and Leakey grantee Cheryl Knott got the idea of collecting urine to analyze

(Opposite) In Borneo, primatologist Cheryl Knott collects orangutan urine beneath a nesting site to analyze female hormone levels. Tracing orangutan fertility cycles has led to vital insights into the survival strategies of these highly endangered apes.

Juni, an adult female, hangs out with her 11-month-old daughter Jane at the Tuanan Research Center in Borneo. Since 2005, Erin Vogel and her colleagues have sought clues here to the puzzle of why orangutans have the longest birth intervals of any mammal.

hormone levels and began tracking the elusive apes in Indonesia's Gunung Palung National Park in the 1990s. "At the time we started here," Knott says, "no one had really worked on hormones in wild apes. People said I was crazy."[81] She collected the urine by spreading plastic sheets under their sleeping nests, then devised an ingenious way to preserve dried samples using filter paper, which she later analyzes in the lab.

When Knott matched the hormone levels of her female orangutans against daily observations of their calorie intake, she found they were responding like women. "When orangutans are losing weight," she says, "during times when there isn't much fruit around to eat, they have lower hormonal levels, which makes it more difficult to get pregnant."[82] Moreover, her urine tests showed that—just like humans but unlike any other great ape—orangutans store fat when food is abundant and burn fat, and ultimately muscle, when times are lean.

Why should such a distinctive, humanlike reproductive cycle have evolved? The forests of Borneo and Sumatra where orangutans live are far less lush and predictable than the African rainforests where chimps and bonobos thrive. The result is a "boom and bust" environment, partly driven by the irregular pattern of the El Nino climate cycle. Long intervals of little or no fruit are punctuated by explosive growth events every four to seven years known as masting, in which many tree species produce massive amounts of fruit simultaneously.

From Knott's work, it is clear that orangutans' reproductive and energetic cycles have evolved in response to extreme ups and downs in food supply. During masting events, the apes gorge on fruit and put on weight; their estrogen levels are high and they mate frequently. During times of scarcity, they survive on less nutritious leaves, stems, and bark, supplemented by termites; they lose weight and their urine contains ketones, a byproduct of fat breakdown, which indicates that they are burning off their fat reserves. When nitrogen isotopes appear in the urine, it is a signal that they are tapping into muscle protein, as humans do when they are on the brink of starvation or suffering from anorexia.

Despite long periods of deprivation, however, orangutans manage to get by—or so Erin Vogel concludes from more than a decade of fieldwork in Borneo's Tuanan forest. "We find no evidence that females stop reproducing," she says, "and infant survivorship remains high."[83] In fact, as in humans, calorie-restricted diets often appear to have a positive health impact, with less damage to cells from oxidative stress, or free radicals. Such similarities, including their ability to store fat, "make orangutans a very interesting model for understanding our current obesity crisis," says Vogel.[84] The similarities also imply that our ancient ancestors, too, may have adapted to the same kind of boom-and-bust extremes that orangutans cope with in Indonesia.

Meanwhile, Borneo has lost an estimated 100,000 orangutans in the first 16 years of the 21st century. If this assessment is correct, the loss represents more than the total remaining population of this graceful and highly intelligent ape. With record rates of deforestation due to palm oil development, they now face vanishing or impoverished habitats that no amount of fine-tuned survival mechanisms can overcome.

AMBUSH AT THE LAKE

Children's books about the past often feature paintings of spear-wielding "cavemen" fighting with formidable beasts such as mammoths and saber-tooth cats. For our early ancestors, the advent of effective weapons must have been crucial for bringing down big game and keeping angry beasts at a safe distance. But when did our forebears acquire them, together with the cooperative tactics necessary to ensure a successful hunt? And when were crude thrusting spears superseded by efficient, long-distance projectiles? A spectacular discovery in Germany stirred up these questions by revealing extraordinarily preserved traces of hunting operations from a remote Ice Age era.

The story begins in November 1995, when archaeologist Hartmut Thieme took a group of colleagues to a rescue excavation he was directing at an open-cast coal mine near the town of Schöningen, about fifty miles from Hanover, Germany. Many of them were skeptical about his claim to have discovered a site where prehistoric hunters had surprised dozens of horses in a highly coordinated effort. According to Thieme, on a dry fall day some 400,000 years ago, a group of hunters had driven a herd of horses into the shallows of a lake and brought them down with wooden throwing spears, or javelins. The ambush yielded a vast bounty of horsemeat, marrow, and hides, which was processed at the lakeside by a large community that lived nearby. They built fires and probably smoked and dried the leftovers for the winter. They also left some of the spears behind at the lake as an offering to the forces they believed had guided them in the hunt.

For Thieme's visitors, this colorful scenario far exceeded their assumptions about the mental capabilities of the humans who were around at this time—either early Neanderthals or *Homo heidelbergensis*. These ancestral Europeans were thought more likely to have engaged in opportunistic hunting or even scavenging rather than coordinated mass kills. Wooden weapons from this era were almost never preserved but it was assumed that they were thrusting spears rather than projectiles. The visiting archaeologists were, therefore, stunned to see a huge array of butchered horse bones, hearths, stone and bone tool fragments, and several miraculously preserved, long, and well-crafted wooden spears, all lying on the ancient lakeshore surface that Thieme had exposed. "The mood on this day can only be described as euphoric," the archaeologists remembered, "as all present quickly realized that they were witnessing a discovery unrivaled in the annals of archaeology."[85]

It was the spears, above all, that commanded wonder. Thieme's team eventually recovered around ten of them, including four intact examples over six feet long. They were "skillfully made,

(Opposite) One of the 300,000-year-old spears excavated at Schöningen, Germany. Carefully crafted from spruce, the spear lies beside the skull and other bones of a wild horse. Experiments with replicas have confirmed their effectiveness as long-range javelins.

with planning and forethought," notes anthropologist Chris Stringer.[86] The material was mostly hard 50-year-old spruce, which was not growing anywhere near the ancient lake. Stone tools had been used to strip them of bark and knots. Thieme observed that their main mass is concentrated toward the front third of the spear, a profile resembling that of today's tournament javelins. In recent experiments with accurately made replicas, skilled throwers were able to hit a target at up to around 60 feet with sufficient impact to kill large prey.

The Schöningen weapons are not the earliest known spears; in South Africa, stone points suitable for hafting to wooden spears and with impact damage on their ends have been dated back to half a million years. Some experts argue that the ability to throw hard and accurately—whether it was rocks or spears—was a vital survival skill, and that physical adaptations for powerful throwing appear far back in the fossil record. They point to several key anatomical changes, notably the ability of shoulder muscles to store and release huge amounts of energy, that first come together in the 1.6 million-year-old skeleton of Turkana Boy (see p. 122).

An opposing view is that the regular use of long-distance projectiles was only developed much later by Ice Age modern humans, long after the time of Schöningen. This view is based on distinctive changes in the shape of the shoulder blade and upper arm bone, which seem related to accurate, high-speed throwing.

Thieme's dramatic interpretation of the site has been questioned. Recent excavations have redated the spear layer to 300,000 years, while the evidence of fires turned out to be natural iron stains in the soil. The number of slaughtered horses is now put at fifty, and they were not all killed at the same time. Instead, the study of isotopes in their teeth, and the wear on the teeth's surfaces, show that the horses came from different places and died at different times, the result of many small-scale hunts over decades or even centuries. Such events would have required fewer hunters and far less coordination than a huge ambush.

Even if these later findings dim some of the initial luster of the Schöningen discovery, the spears remain a marvel of preservation and testify to remarkable forethought and craftsmanship. It took planning, cooperation, and perhaps language to create the spears, anticipate the behavior of the horses, and surprise and drive them into the shallows of the lake. Schöningen helped put to rest the notion that their makers were ineffective, marginal predators and, instead, encouraged us to see them as skilled and strategic hunters.

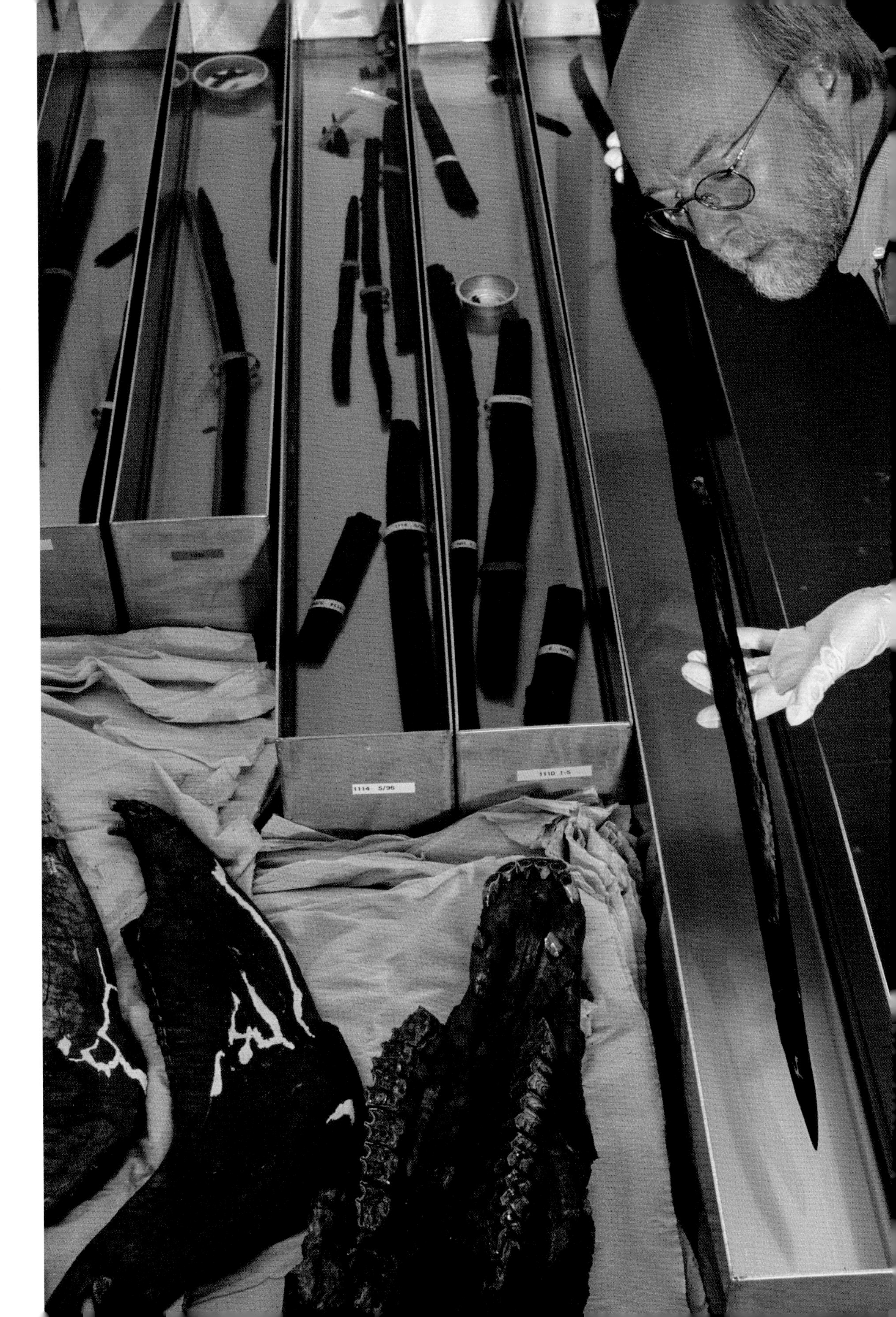

Archaeologist Hartmut Thieme examines the Schöningen spears that his team excavated in 1995.

OUR AFRICAN MOTHERS

In 1987, a paper in the journal *Nature* presented a revolutionary application of the newfound tools of DNA analysis to probe the origins of modern humans. Hugely controversial and easily misunderstood when it was first published, the paper's conclusion boiled down to the idea that part of the genetic heritage of all today's women can be traced back to a single female ancestor who lived in Africa some 200,000 years ago. An accompanying news article in *Nature* was titled "Out of the Garden of Eden" and, soon, the media was hailing this ancestor as our "African Eve." *Newsweek* magazine breathlessly imagined her as a "dark-haired, black-skinned woman, roaming a hot savanna in search of food," who is "as muscular as Martina Navratilova, maybe stronger" and "might have torn animals apart with her hands."[87]

The study was conducted by two University of California, Berkeley graduate students, Rebecca Cann and Mark Stoneking, and their professor, Allan Wilson, who promoted the idea that slight variations in DNA can serve as a kind of molecular clock. These variations—"copying mistakes" or mutations—accumulate at a known rate over time. If you compare DNA variations between individuals from two living populations or species, you can estimate when they last branched off from a common ancestor. The more similar their patterns of DNA, the more recently they shared an ancestor.

The Berkeley team focused on what is known as mitochondrial DNA. Mitochondria are tiny structures within each cell that convert oxygen and nutrients into energy. Unlike nuclear DNA, which plays the central role in creating our bodies and identity and is a mixture of both parents' genes, mitochondrial DNA is only passed down from mother to child through the female line. Since there is no recombination or mixing of genes, mutations or copying mistakes tend to persist from one generation to the next, forming markers that allow scientists to trace female lineages back to a specific ancestor. After the 1987 study, mitochondrial DNA—or "mtDNA"—was recognized as a powerful way to pin down the origins and migrations of populations over time.

By today's standards, the methods available to the Berkeley team were primitive. While a cheek swab is all that is needed to take a DNA sample now, Cann had to persuade 147 mothers from around the world to donate their placenta. To analyze the DNA data, the team laboriously wrestled with punch cards, mainframe computers, and reams of printed paper output.

With the computer's help, the team plotted out over 100 distinctive mitochondrial DNA variations, or lineages, arranged in a branching tree that connected the most similar together over time. The results were a revelation. "When you put the big picture together," says Cann, "you understood that the very base of the tree might be as much as 200,000 years old, and it was overwhelmingly African. The inference was that one woman at the base of that mtDNA tree was the universal lucky mother of all modern people, the one woman with an unbroken line of maternally inherited genes that came down to us today."[88]

Thanks partly to the Garden of Eden metaphors, this striking result stirred up much confusion. The Cann team had *not*

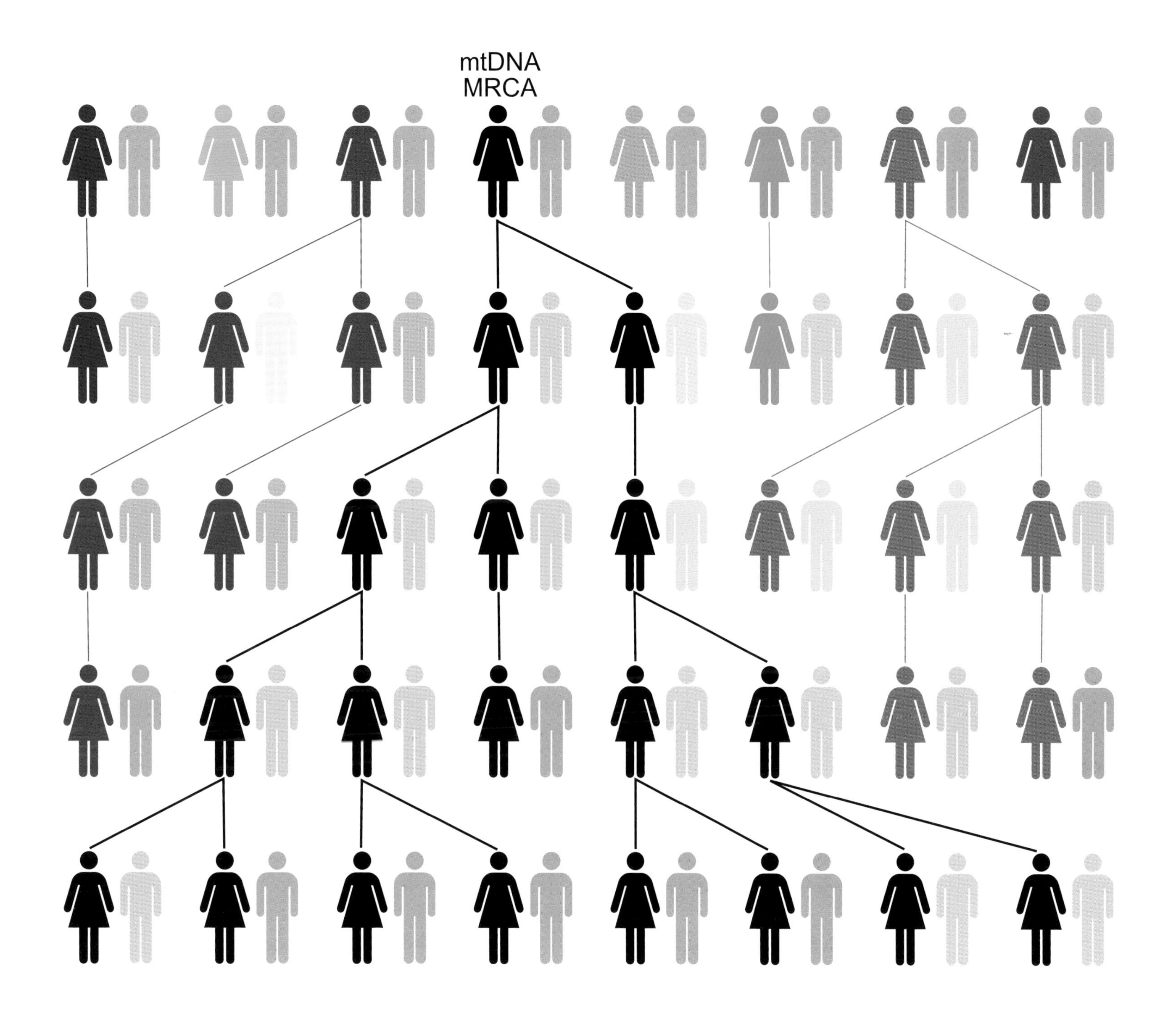

This diagram shows how a single line of mitochondrial DNA (in black) came to be passed down to all women today, originating from a "Mitochondrial Eve" who lived in Africa some 200,000 years ago.

(Opposite) A grandmother of the San people of South Africa holds a child. Despite the limitations of the "Mitochondrial Eve" study, subsequent analysis of the genetic diversity present in today's Africans confirms that modern humans originated there.

identified humanity's first mother, nor was "African Eve" the only woman alive at the time she lived; it meant simply that all the mtDNA present in women today traces back to that one individual. "She was one individual out of thousands of ancestors," explains anthropologist John Relethford, "but she was the only one in that generation who contributed to the mtDNA of everyone alive today."[89]

The relatively recent and exclusively African origin of mtDNA sent shockwaves through the academic community. At the time, two competing models explained how modern humans had emerged around the globe. The "Multiregional" view saw human populations as a single continuous lineage that evolved gradually across Asia and Europe as well as Africa. The "Out of Africa" view pictured an abrupt emergence of modern humans in Africa around 100,000 years ago. Then, endowed with superior traits and technology, a small group left Africa and spread rapidly across Europe and Asia, driving disadvantaged species such as the Neanderthals and *Homo erectus* into extinction. By locating the origin of all our modern mtDNA in Africa a mere 200,000 years ago, the "Eve" study decisively favored the "Out of Africa" model.

Although the analysis was groundbreaking, mitochondrial DNA represents a tiny fraction of our genes. The team had illuminated only a single genetic pathway among tens of thousands that stretch back into our ancestral past. "Each gene or DNA sequence can have a different history," says Relethford, "and we can trace our ancestry back to multiple different common ancestors, potentially living at different times and places."[90] By 2006, robots that automated the task of reading DNA made it affordable to sequence a person's entire genome. As a result, scientists could now reconstruct the history of those tens of thousands of nuclear genes. Most of their gene trees are rooted in Africa, confirming that modern humans did indeed originate there. But the timing of each gene tree's split from its last common ancestor varies hugely. Some date back nearly 2 million years, while most of them are considerably older than the mtDNA "Eve" date of 200,000. Genetic evidence now points to many different populations across Africa contributing to the gradual birth of our species over an immense timespan, rather than a single and fairly recent event.

As a pathbreaking early exploration, Mitochondrial Eve struck a deep popular chord and told a bold, appealingly straightforward story about our ancestral roots. Whatever its limitations, the study provided a model for investigating the genetic record of our species, and mitochondrial DNA is still widely used to track the origins and migrations of today's populations around the globe.

THE EARLIEST "US"

Today's Sahara desert straddles the African continent, a forbidding barrier almost the size of the United States. But some 300,000 years ago, during an episode of milder climate, the Sahara was "green"—covered with abundant grasslands and lakes that supported herds of gazelles, zebras, wildebeest, and buffalos. A group of ancestral humans hunted these beasts at a site known as Jebel Irhoud, some 70 miles from Marrakesh in Morocco. They had mastered fire and were skilled flint toolmakers. The cave in which they sheltered preserved the enigmatic skull remains of more than ten of these individuals, rare evidence of North Africa's neglected human fossil record. A new study of these skulls in 2017 reinvigorated an old debate about when, where, and how the first anatomically modern humans emerged across the continent.

In the debris of a collapsed ancient cave, a stunningly complete skull was discovered by accident during mining operations at Jebel Irhoud in 1961. At first, it was thought to belong to a Neanderthal, perhaps dating to around 40,000 years ago. When that date was later revised to 160,000 years, the skull and other human remains found at the site became a puzzling anomaly. Their blend of advanced and primitive features suggested that they might be the result of contact between modern humans and Neanderthals, or perhaps a lingering remnant of an even more ancient population, isolated in a North African backwater.

Paleoanthropologist Jean-Jacques Hublin, long fascinated by North Africa's past, was determined to solve the puzzle. It took two decades to overcome political and funding hurdles to gain access to the site, but, eventually, he was rewarded by finding three new skulls and two jaws, and a startling new date of around 300,000 years ago. This was some 100,000 years older than East African fossils that seemed to represent the earliest examples of anatomically modern traits.

(Opposite and above) In 2004, excavations at the Jebel Irhoud quarry in Morocco recovered teeth, jaws, partial skulls, and limb bones from at least five individuals, along with burned stone tools indicating control of fire. The claim that the 300,000-year-old fossils represent the earliest *Homo sapiens* rekindled a debate on the origins of the first anatomically modern humans in Africa.

Based on CT x-ray scans and hundreds of 3D measurements of the skulls, Hublin's analysis drew attention to their mosaic of features. Viewed from the front, their small faces and lower jaw and teeth resemble those of modern humans. "The face is that of somebody you could come across in the Metro," Hublin says.[91] But viewed in profile, their elongated braincases, sharply angled at the rear, are quite different from the globular skulls we have today. Hublin argues that this distinctive skull shape reflects a less advanced organization of the brain, and that genetically-driven shifts in brain structure and connectivity, crucial for fully modern cognition, had yet to emerge. We might recognize the familiar look of the Jebel Irhoud faces, but not their

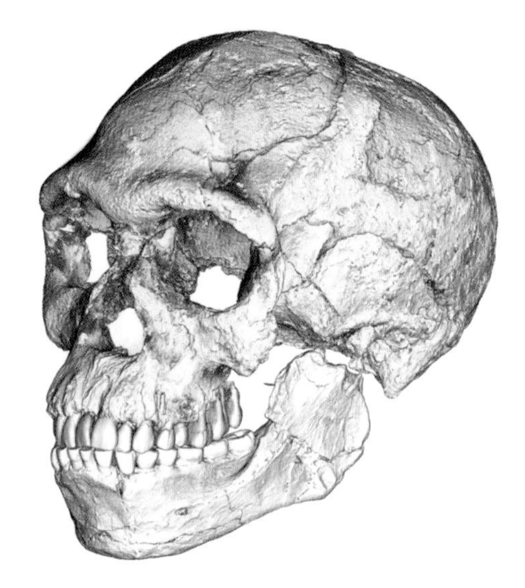
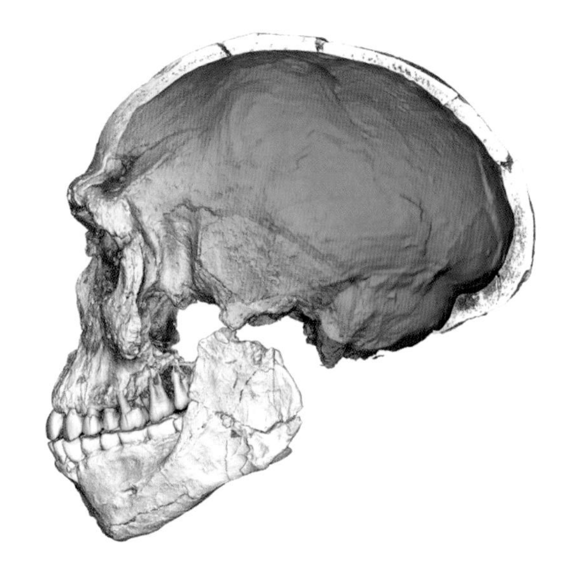

A composite computer reconstruction of a Jebel Irhoud skull based on CT scans of several fossils from the site. Viewed from the front, the face looks quite modern, but in profile, its flat, elongated shape is more typical of earlier, archaic ancestors. (Opposite) A reconstruction based on the Jebel Irhoud I skull.

thinking and behavior. Hublin adds, "It's vain to hope to find humans that would be called 'modern' 300,000 years ago, and that would be fully modern in their anatomy and behavior. They are in the lineage of modern humans, but they are different."[92]

Nevertheless, Hublin has also argued that the Jebel Irhoud skulls represent "early modern humans"[93] and that they lie "at the very root of our species, the oldest *H. sapiens* ever found in Africa or elsewhere."[94] That claim raises "major questions about what features define our species," notes paleoanthropologist Marta Mirazón Lahr.[95] "Is it the globular skull, with its implications for brain reorganization, that makes a fossil *H. sapiens*? If so, the Irhoud population represents our close cousins," not our actual ancestors. But if small faces and jaw shapes are held to be crucial traits, then the skulls could, indeed, belong to our species.

The debate is complicated by the remarkable variety of African fossils from this era, which are "much more diverse than skulls taken from any single population today," according to paleoanthropologist John Hawks. "They were relatives," he says, "but they were not all modern."[96] For instance, two fossil skulls unearthed by Richard Leakey at Omo Kibish in Ethiopia, are both around 195,000 years old, yet are quite different: one looks very modern, with a short, broad face, high forehead, and rounded skull bones, while the other, far more rugged, has a similar elongated shape to the skulls from Jebel Irhoud. So was there one, highly variable population in a single part of Africa that gave rise to us? Or did modern traits emerge in many places and commingling populations?

The Jebel Irhoud fossils show that such traits could arise in a region far removed from the traditional cradle of our species in eastern and southern Africa. During the favorable conditions of the "Green Sahara," human communities were no longer confined by desert barriers and so they could travel and exchange genes across long distances. At other times, rapid climate shifts and geographical barriers led to isolation and a patchwork of distinct populations. Most scientists now view our modern features as springing from the changing interactions of many peoples across Africa over at least half a million years. As archaeologist Eleanor Scerri puts it, "The evolution of human populations in Africa was multi-regional. Our ancestry was multi-ethnic. And the evolution of our material culture was multi-cultural. We need to look at all regions of Africa to understand human evolution."[97]

THE ORIGINAL SOCIAL NETWORK

At Easter 1942, Louis and Mary Leakey loaded up their two old Chevrolets and took off for a weekend excursion to the Kenyan Rift Valley. They were following up a report that stone tools had been found on the fringes of a dried-up ancient lake at Olorgesailie, 40 miles southwest of Nairobi. After arriving and trekking through the bush, both Louis and Mary suddenly came upon an extraordinary sight: vast arrays of hundreds of hand axes lying on the surface, "looking as if they had only just been abandoned by their makers," as Mary wrote.[98] They had stumbled across a major site, parts of which would eventually be dated back to over a million years and would prove to contain a wealth of clues about the emergence of modern human behavior.

For the next decade, Mary periodically dug at Olorgesailie but "never did make much real progress," she admitted,[99] due to ancient landscape and climate shifts that made the geology of the site hard to disentangle. In 1985, Mary visited another site in the Rift Valley where Rick Potts, a newly hired scientist at the Smithsonian's National Museum of Natural History, was digging. She mentioned that Olorgesailie "had never been done properly," and could benefit from new techniques of excavation and dating. Her suggestion led to an ambitious, three-decade joint project of the Smithsonian and the National Museums of Kenya. Led by Potts, the international team has recovered

Paleoanthropologist Rick Potts contemplates an accumulation of 700,000-year-old hand axes at Olorgesailie in Kenya. For three decades, Potts's team investigated the shifts between climate, ecology, and human activities at the site.

a phenomenally detailed record of shifting patterns in Olorgesailie's ancient environment, climate, and settlement.

Based on multiple digs, the team has traced over half a million years of activity at the site by the hand axe makers. They crafted these relatively simple tools from local volcanic rocks and used them to butcher large mammals such as elephants, zebra, and pigs that were drawn to the ancient lake shore. Then landscape upheavals led to the draining of the lake and, around 400,000 years ago, abrupt swings between wet and dry climates began. Many species of large grazing animals went extinct and once predictable food sources grew scarce.

Under this pressure, a remarkable change in the human cultural pattern at Olorgesailie took place. Hand axes disappeared and a completely different range of smaller, more finely crafted and specialized stone tools took over. These included small, triangular points that could be hafted and turned into projectile weapons, making hunting more effective. The blades and flakes used to make the points were struck off from carefully prepared cores, a technique that demanded more strategic thinking and skill than the earlier hand axes.

With this growth in technical skills came an expanded choice of stone. As well as the local lava rocks exploited for hand axes, the new toolmakers used obsidian—a black volcanic glass that could be sharpened as finely as today's surgical steel. A team led by Potts's colleague, archaeologist Alison Brooks, used chemical analysis to pin down the obsidian's source. Remarkably, it had come from five different directions, ranging from 15 to 55 miles away. Huge quantities of obsidian flakes at Olorgesailie suggested that raw stone had been imported in bulk and then shaped on the spot. All the evidence pointed to a striking conclusion: the volcanic glass had been traded across a long-distance social network. This implies an essentially modern ability to plan, coordinate, and communicate with people outside the immediate community.

An even bigger surprise were dozens of dark-colored lumps. Potts recalls one of his diggers picking one up and exclaiming when he found that it left black marks on his skin. Under the microscope, these calcite lumps proved to be streaked with manganese oxide, a mineral often used in prehistory as black pigment. They had been brought to Olorgesailie from at least 15 miles away. Other red-colored lumps, some bearing scratch marks, were red ocher or hematite, another mineral commonly used for pigment. "We don't know what the coloring was used on," Potts says, "but coloring is often taken by archaeologists as the root of complex symbolic communication. Just as color is used today in clothing or flags to express identity, these pigments may have helped people communicate membership in alliances and maintain ties with distant groups."[100]

Who traded the stone and painted with the pigments? Similar advanced tools were found at Jebel Irhoud Cave in Morocco, alongside human fossils that show a mixture of archaic and modern-looking features (see p. 155). At 320,000 or more years ago, the dating of the innovative behavior at Olorgesailie is a good match for that of the North African skulls. It is tempting, then, to link the emergence of anatomically modern traits with the advent of social networks and symbolic expression. "What I think we're seeing at Olorgesailie," reflects Potts, "is a change to the way people think today. It's a change from an immediate, localized way of thought to one in which you make mental models of the landscape and the social groups you might run into tens or hundreds of miles away. And that's a very different mentality that's right at the root of *Homo sapiens*' ability to adapt to change."[101]

THE PIT OF THE BONES

For young boys from the village of Ibeas de Juarros, in northern Spain, venturing into the nearby Sierra de Atapuerca used to be a traditional rite of passage. Entering a limestone cavern called Cueva Mayor, the boys would walk, wriggle, and crawl for a third of a mile until they reached a 43-foot vertical shaft known as the Sima de los Huesos, or Pit of the Bones. Then they would clamber down the shaft to reach the Pit's sloping floor, which led to a cramped, dead-end chamber. There, in the clammy mud, they would dig up the teeth of extinct cave bears to take back to the village and impress their girlfriends. Centuries of all this tooth hunting had so thoroughly churned up the mud that the Pit attracted little attention until a graduate student spotted a fossil human jaw there in 1976. In the 1980s, a team of archaeologists began an investigation, which first required backpacking tons of wet clay out of the Pit. Improbably, the tiny area of the dead-end chamber—no bigger than an average city parking space—turned out to conceal Europe's single richest deposit of fossil humans.

Three decades of digging have yielded a trove of nearly 7,000 fossils representing at least 28 individuals. They appear to belong to one population and they provide an unprecedented "snapshot" of a single prehistoric people, who lived around 430,000 years ago. Forensic study of the bones reveals these people to have been robust, suggesting physically demanding lives, with both sexes averaging a height of around 5 feet 6 inches. Their brains were smaller than the modern average. Most of the bones belonged to teenagers; only three adults survived beyond the age of 30. The Pit's impressive array of 17 skulls—some of the most intact ever found in Europe—exhibit distinctive features such as thick brow ridges arching over each eye and rugged lower jaws with small molar teeth, all foreshadowing the characteristic appearance of the Neanderthals. A similar conclusion springs from an extraordinary feat of DNA analysis, one of the oldest sequences yet retrieved from ancient human bone. Based on samples from a tooth and a femur, geneticist Matthias Meyer laboriously reconstructed DNA fragments of only 20–30 base pairs or nucleotides at a time. Eventually, he was able to piece together over 1 million base pairs. This was a tiny fraction of the 3 billion base pairs of the nuclear genome, yet it was enough to confirm a strong link between the Sima people and the Neanderthal lineage.

These "pre-Neanderthals" faced many health challenges. The best-preserved skull of all, nicknamed Miguelón by the archaeologists, belonged to a large male. His heavily worn teeth suggested he had used them as a "third hand," probably to grip hides while defleshing animals. When a tooth broke, an infection spread into his upper jaw and up to his eye socket, likely

(Opposite) Archaeologists toil in the narrow shaft of the Sima de los Huesos in 2017. In this hidden pit at Cueva Mayor Cave in northern Spain, at least two dozen bodies of Neanderthal ancestors were dragged and deposited over 400,000 years ago.

(Opposite) At least 17 skulls from Sima de los Huesos foreshadow the anatomy of the ruggedly built Neanderthals, who appear in Europe not long afterwards. (Above) Nicknamed "Excalibur," this unusual hand axe made of exotic pink quartzite was the sole tool found in the burial pit, perhaps an offering to the dead.

leading to a painful death at age 35–40. Another adult male, Agamemnón, was probably deaf from bony growths that invaded his ear canals. But the most remarkable case was Benjamina, a little girl whose head was abnormally lopsided due to a condition known as craniosynostosis, in which a skull plate closes prematurely during infancy. She undoubtedly suffered mental disability, yet she survived until the age of around nine. The presence of these diseased and disadvantaged individuals strongly suggests a measure of social care and empathy among the prehistoric community at Atapuerca.

But could this sense of empathy have extended to a concern for the dead? The idea that the bodies were deliberately dropped into the Pit of Bones is controversial; even more so, is the theory that this was a ritual act of intentional, rather than casual, disposal. Yet after decades of study, the archaeologists have examined every possibility and one by one eliminated other plausible explanations for how the bodies could have found their way into the Pit.

Clearly, some had died from fatal disease before their bones ended up there. No evidence exists that they were transported there by water or other natural means. Nor was the Pit a carnivore den, although bears, lions, and foxes occasionally tumbled into the shaft. The tiny chamber was not a living site, as no tools have been found with the bodies, with one tantalizing exception. The sole artifact found in the Pit is a hand axe, crafted of beautiful and unusual pink quartzite, which the team named "Excalibur" after King Arthur's legendary sword. It was made from "a raw material that was quite special and has no traces of being used," says María Martinón-Torres, one of the Atapuerca paleoanthropologists and a Leakey Foundation grantee. "After analyzing all the possibilities we think we may have one of the earliest examples of an intentional accumulation or disposal of bodies, with this very enigmatic hand axe deposited with them. So we may have one of the first instances of a funerary practice, something we will continue to explore."[102]

RESOURCEFUL "BRUTES": THE PEOPLE OF THE ROCK

The soaring cliffs of the Rock of Gibraltar have attracted humans for over 100,000 years. A string of caves fronting the ocean was home to prehistoric foragers, who fished for bream and tuna, collected mussels, snared rabbits and pigeons, and hunted ibex and red deer in the patchy woodland bordering the sea. They also left markings on a cave floor that may have held symbolic meaning, and they captured ravens and eagles, probably to adorn themselves with dark wing feathers. In nearly every respect, what we know about their lifestyle matches the range of behavior and abilities typical of *Homo sapiens*, the modern humans who had begun entering Europe around 45,000 years ago. Yet the occupants of Gibraltar were not modern humans, but Neanderthals, long imagined in popular culture as dimwitted sub-humans. Discoveries from two Gibraltar caves over the last quarter-century show how much we have underestimated the minds and abilities of our extinct cousins, who had populated wide areas of Europe and Central Asia for over 300,000 years before the arrival of modern humans.

Gibraltar was where the first adult Neanderthal skull was discovered in 1848. However, it was not recognized as an extinct human until 1864, the same year that geologist William King coined the species name *Homo neanderthalensis* after studying fossil bones found in 1856 in the Neander Valley, Germany, a discovery that had attracted far more attention. King believed that the ancient human's "thoughts and desires ... never soared beyond those of a brute,"[103] setting the tone for a century and a half of Neanderthal stereotypes.

Since 1994, Clive and Geraldine Finlayson of the Gibraltar Museum have led the investigation of Gorham and Vanguard caves, which were occupied by Neanderthals for more than 100,000 years. Both caves are filled with layers of windblown dune sand that conceal hearths and living areas preserved in fine-scaled detail—sometimes, even traces of individual meals. The iconic Ice Age big game of mammoth, bison, and reindeer never made it as far south as Gibraltar, where milder climate conditions prevailed. Instead, the Neanderthals hunted deer and ibex, fished, trapped small game, gathered pine nuts, and even ate seals and dolphins. "The picture that is emerging is that Neanderthals had a diverse larder outside their cave window and they were exploiting all these things," says Clive Finlayson.

Among their quarry, the Gibraltar Neanderthals trapped more than 150 species of birds, notably raptors and corvids (such as ravens). Microscopic examination of cutmarks on the bones of these birds of prey indicates that they were not being butchered for their meat but to extract their plumage—specifically, their glossy dark wing feathers. The cutmark patterns also

Excavations at Gorham's Cave on the Rock of Gibraltar have yielded surprising insights into Neanderthal behavior and lifeways. The cave dwellers subsisted on fish, mollusks, seals, tortoise, rabbit, and red deer. They also used large raptor feathers either for rituals or personal decoration.

suggest that the Neanderthals were separating the bones from the feathers while leaving the skin intact, potentially creating an impressive cape or headpiece that they could have worn. The Gibraltar team found the same preference for raptors and corvids in finds from many other Neanderthal sites across Europe and Asia. In Croatia's Krapina Cave, for example, eight talons from white-tailed eagles were laboriously extracted and polished to form a bracelet or necklace that dates back to 130,000 years ago.

As well as wearing feathers and claws, Neanderthals likely painted their bodies and other surfaces. Digs at many European sites have turned up lumps of the mineral manganese dioxide, widely used in prehistory as a black pigment. At the Pech de l'Azé rock shelter in the Dordogne, more than half of at least 500 pieces of the mineral showed traces of polish and wear, and some had been modified into pointed, crayon-like shapes.

If Neanderthals were decorating themselves, what of the cave walls in which they sheltered? In 2014, the Gibraltar team reported the discovery of a lattice-like carving on a flat bedrock surface at Gorham's Cave, dating to at least 39,000 years ago. Nicknamed the "hashtag," experiments showed that the pattern was not a casual doodle but a laborious engraving, probably requiring at least 200–300 repetitive tracings by a flint tool.

Still more remarkable news surfaced from a 2018 study of three painted cave sites in Spain. A team led by archaeologist Alistair Pike applied a newly refined technique to dating precipitated limestone or flowstone. Thin crusts of these minerals slowly creep across cave walls, covering up ancient artwork in the process. Measuring the amount of radioactive uranium and thorium in the flowstone can give a minimum date for the underlying artwork. When Pike's team applied the technique to red-painted designs in the three different caves, the results were startling. All three had age ranges suggesting that they must be more than 65,000 years old, made long before any modern humans were in Europe. The red motifs range from a ladder-like pattern to the ghostly outline of a "negative hand" impression, made by blowing pigment around a hand pressed against a cave wall. The accuracy of the flowstone dating technique is controversial, but if the age is correct, it would mean that Neanderthals, not modern humans, were responsible for a previously unsuspected early stage of cave art.

This tantalizing recent evidence of self-expression challenges long-held assumptions about Neanderthal inferiority. Some experts hold to the belief that their brains were wired in fundamentally different ways to our own. For example, paleoanthropologist Ian Tattersall points to the absence of art and personal ornaments from the vast majority of Neanderthal sites, and the lack of innovation in their stone tools during most of their existence in Europe. He concludes that "behaviorally ... the Neanderthals were simply doing what their predecessors had done, if apparently better. In other words, they were like their ancestors, only more so. We are not. We are symbolic."[104] Yet, at least toward the end of their long sojourn in Europe, it looks as if Neanderthal culture and capabilities were steadily advancing, calling into question the notion of some fundamental biological difference. Even if their brains *were* wired differently, it would surely be wrong to view them as inferior or unable to thrive. Archaeologist Paola Villa notes that "Neanderthals lived for 350,000 years under various climate conditions, longer than modern humans have been around. They were not brutes and we are now seeing how adaptable and exceptional they were."[105]

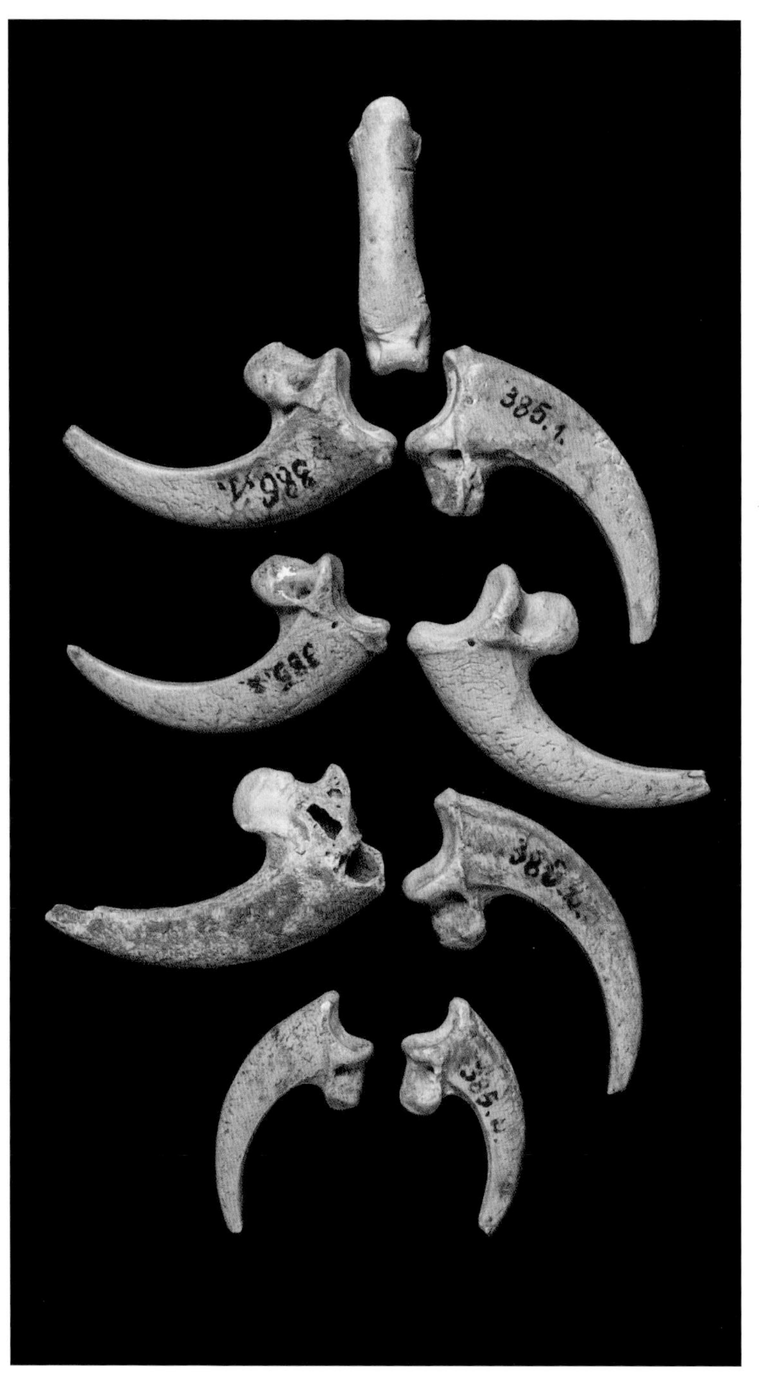

(Above) The so-called "hashtag" is an abstract design carved on a flat surface inside Gorham's Cave and associated with a Neanderthal occupation level dated to at least 39,000 years ago. (Right) A 130,000-year-old necklace of eagle talons from the Krapina rock shelter in Croatia, the earliest evidence for the symbolic or decorative use of raptor feathers by Neanderthals.

CRACKING THE NEANDERTHAL CODE

In June 2006, Swedish geneticist Svante Pååbo traveled to Zagreb, Croatia, on a momentous mission: to find Neanderthal bones with enough preserved DNA to recover the first complete Neanderthal genome. He had already run promising tests on a bone from Vindija Cave, a spacious cavern 50 miles from Zagreb, where Neanderthal remains were excavated in the 1970s and 1980s. Most of those bones were deliberately broken and crushed, and bore cut-marks identical to traces left behind by everyday meat butchering. This strong evidence for cannibalism has been found at over half-a-dozen other Neanderthal sites, although whether it was driven by starvation, ritual practices, or a casual attitude to the dead is much debated.

Whatever motivated it, the intentional breakage of the bones was a boon to Pååbo. If the Neanderthals had buried intact corpses in the cave, they would have taken months to rot, giving microbes more opportunity to invade the bone cells and degrade DNA. Yet bacterial damage was only one of the hurdles that Pååbo and his team faced. Ancient DNA also deteriorates if it is subjected to too much heat, moisture, or acid soils. Another huge issue was contamination by the DNA of modern people, including Pååbo's own team. They had to build an ultra-sterile lab resembling the "clean rooms" where microchips are manufactured.

(Opposite) A Neanderthal bone fragment is drilled to extract DNA at the Max Planck Institute for Evolutionary Biology in Leipzig, where many of the breakthroughs in analyzing ancient DNA were pioneered.

In this new facility, located at the Max Planck Institute in Germany, the team drilled samples from the Vindija bones. Most yielded negligible traces of Neanderthal DNA, but three bones proved to contain more than 1 percent and, in one case, nearly 3 percent. It was a start, but would there be enough to retrieve the entire nuclear genome?

Before 2006, the technology to answer this question barely existed. The first modern human genome had been published only five years earlier, and that took more than a decade of international collaboration to accomplish. New "high-throughput" sequencing machines, which could identify millions of DNA's nucleotide building blocks simultaneously rather than one at a time, made Pååbo's project look feasible. But the team still faced Herculean challenges, such as devising new statistical techniques and chemical probes to weed out errors and contamination from genuine fragments of ancient DNA.

Finally, in 2010, they were ready to announce their findings to the world. The first revelation in the sequenced genome was that, overall, our genetic identity is extremely close to that of Neanderthals—about 99.7 percent of our genes are identical. Still, there are differences, mostly due to random changes (or mutations) in our DNA since the Neanderthal branch split from the one that eventually led to modern humans. These differences allowed Pååbo's team to estimate that our last common ancestor had lived at least 300,000 years ago.

But the real shock came when they compared the Neanderthal genome to that of living Africans, Europeans, and Asians. Most experts, including Pååbo, subscribed to a simple version of the "Out-of-Africa" theory. In this scenario, modern humans began their major exodus from Africa around 60,000 years ago. As they spread into the Neanderthals' homelands in Western Asia and Europe, they totally replaced them without any significant interbreeding. If this was indeed true, then the Neanderthal genome would be equally close to that of all populations around the world today. Instead, the newly decoded data revealed something different: Neanderthals are *more* closely related to people living outside Africa than they are to Africans.

The most logical explanation is that some modern humans interbred with Neanderthals shortly after leaving Africa. In fact, the genetic relics of these encounters are still present in our bodies. Although most people in Europe and Western Asia carry around only 2 percent of the Neanderthal genome, each individual carries different parts of it, so that a total of about 20–30% of the genome survives in today's non-African populations. While Neanderthals died out as a distinct population, their genetic legacy lives on in many of us.

Many researchers are now investigating how versions of the genes we inherited from Neanderthals could be affecting our health today. An intriguing finding is that some of the genes in our HLA system, which controls the immune response of white blood cells, were passed on by Neanderthals. As modern humans moved into Eurasia, they faced threats from unfamiliar pests and pathogens. Since Neanderthals had already been living there for hundreds of thousands of years, their immune systems were better armed and, evidently, their HLA and other genes involved in immunity offered a significant advantage to our ancestors. Yet our Neanderthal heritage has negative consequences, too. Variations of their genes affect our risk of developing type two diabetes, the spread of belly fat, and the buildup of LDL (the so-called "bad" cholesterol linked to heart attacks). In 2020, an even more striking discovery revealed that a segment of DNA linked to a three-fold risk of developing severe COVID-19—thought to be a major factor in affecting mortality in the epidemic—came from Neanderthals. ("I almost fell off my chair," said co-investigator Max Zeberg when he first saw the analysis, "because the segment of DNA was exactly the same as in the Neanderthal genome."[106]) While this and other Neanderthal genes may have tragic consequences today, they must have conferred advantages to early modern humans as they encountered Neanderthals, perhaps protecting them from other diseases no longer present in our modern world.

The decipherment of the Neanderthal genome was an astonishing achievement, with implications that researchers are still exploring and debating. Looking back at the project, one of Pååbo's collaborators, David Reich, says he continues "to have nightmares that the finding is some kind of mistake. But the data are sternly consistent: the evidence for Neanderthal interbreeding turns out to be everywhere. As we continue to do genetic work, we keep encountering more and more patterns that reflect the extraordinary impact this interbreeding has had on the genomes of people living today."[107]

(Opposite) Trailblazing geneticist Svante Pääbo confronts a reconstructed Neanderthal skull at the Max Planck Institute in Leipzig.

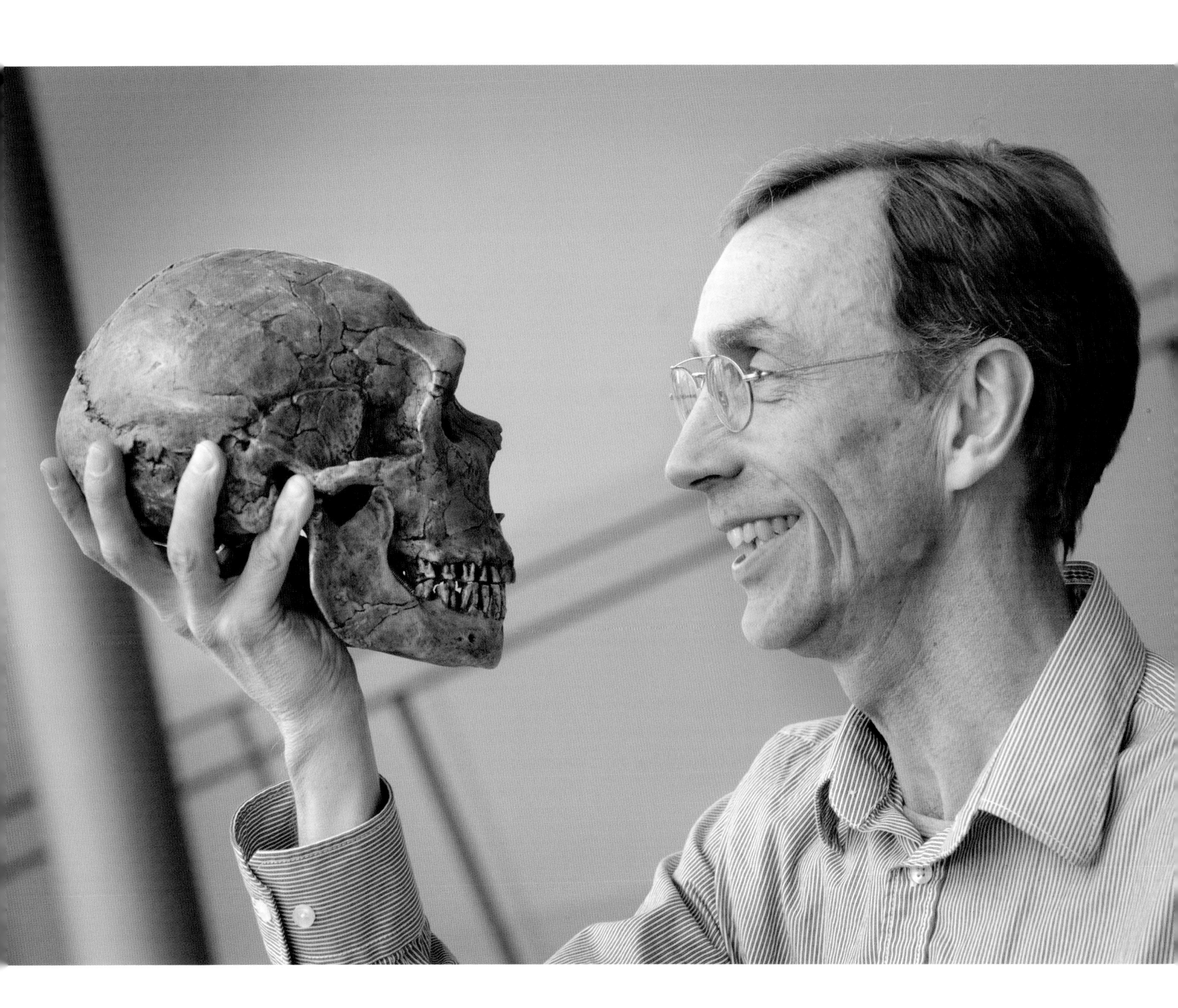

THE FINGER IN THE CAVE

One day in July 2008, archaeologist Alexander Tsybankov came across a tiny bone belonging to a child's pinky finger inside Denisova Cave in the Altai Mountains of southern Siberia. Here, for more than three decades, Russian archaeologists had been digging through layers of recent goat dung to reach soil loaded with ancient bones and stone tools. Tsybankov bagged the insignificant-looking scrap and took it back to camp, little suspecting that it concealed traces of a previously undiscovered population of prehistoric humans that existed for at least 400,000 years before it vanished.

The pinky bone was sent to the Max Planck Institute in Leipzig, where Svante Pääbo's team was deeply engaged in its landmark decoding of the first complete Neanderthal genome. The bone attracted little attention there until 2009, when Pääbo's colleague Johannes Krause decided to extract and sequence its mitochondrial DNA. Since the bone had been dated to 50,000 to 30,000 years ago, he was expecting the DNA to belong either to a Neanderthal or a modern human. Instead, it represented a completely unknown, extinct human. Krause recalls the discovery of this mystery people, later to be known as Denisovans, as "scientifically the most exciting day of my life."[108]

The following year, two fossil Denisovan teeth were found at the cave dig, one of them twice the size of a modern molar—so massive, in fact, that it was at first mistaken for a cave bear's tooth. But it was the child's pinky bone that commanded the Leipzig team's attention, because it preserved a staggering 70 percent of the original ancient DNA, by far the largest amount ever retrieved from a single human bone. (Frigid year-round conditions at Denisova Cave help explain the exceptional preservation.) This sole specimen yielded more of the entire Denisovan genome than they had acquired from *all* their prior work on Neanderthals.

From their analysis, Pääbo's team discovered that the Denisovans had a highly distinctive genetic identity; they had split off from the ancestral population of Neanderthals at least 400,000 years ago, well before many of the Neanderthals' most distinctive traits had begun to emerge. One aspect of Denisovan DNA shared with Neanderthals' was traces of inbreeding, suggesting that they, too, had lived in small, isolated populations. But where, beyond the cool interior of Denisova Cave, had this unknown people lived?

The answer involved tracking down Denisovan DNA among today's populations. The resulting analysis sprang another huge surprise: the highest percentages were in Melanesians and among Aboriginal Australians, in some areas reaching 4–6%, with lesser amounts in eastern Indonesians, southeast Asian islanders, and East Asians. This was a mindboggling pattern.

(Opposite) The spacious main chamber at Denisova Cave in the Altai Mountains of Siberia. A fragment of finger bone and two molars found in a small side chamber led to the momentous announcement of a previously unknown branch of humanity in 2010.

Thousands of miles separate the chilly Altai Mountains from Indonesia's steamy tropical forests, environments requiring utterly different adaptations and lifeways. What is more, Denisovan genes turned up in the extreme west, in mitochondrial DNA extracted from the 430,000-year-old remains found in the "Pit of Bones" at Atapuerca in northern Spain (see p. 162). How did Denisovan people come to span this vast range of Europe, Asia, and southeast Asia?

There have been several competing explanations, but in 2019, a new study of modern genomes from southeast Asia revealed just how complicated our picture of the peopling of ancient Eurasia has become. It now looks as if there were at least five major human populations, all still co-existing as recently as around 70,000 years ago: modern humans, the Neanderthals, and three distinct groups of Denisovans, one in Siberia and the other two in southeast Asia. One of these Asian groups, identified in genomes from Papua New Guinea, differed as much from Siberian Denisovans as it did from Neanderthals. To account for all the details of the genetic data, the Denisovans must have interbred several times with both Neanderthals and modern humans. Moreover, the Denisovans also mixed with members of at least one other "ghost population" in Asia, a mystery group that has yet to be identified with any fossil remains. Its existence was recently detected in the genome of living Asians by an A.I. machine learning program, trained by scientists to run through hundreds of thousands of simulations of demographic and genetic data.

One way to visualize this intricate picture, suggests anthropologist John Hawks, is to think of it like a medieval manuscript, scrawled over and erased many times by monks, with different layers of text partially visible and rubbed out. Similarly, he says, "our ancestry comes from multiple layers of ancient movements that emerged from Africa, became somewhat differentiated from each other, mixed with each other, and then remixed again as later people left Africa. It's a really complicated and interesting scenario and the Denisovan genome demonstrates it more than anything else."[109]

(Opposite) Despite its tiny size, this fragment of pinky finger bone from Denisova Cave contained the biggest proportion of ancient human DNA ever recovered from a sample, enough to identify the human relatives we now call Denisovans.

THE FATE OF THE HYBRIDS

Tens of thousands of years ago, according to landmark genetic evidence, modern humans interbred with both Neanderthals and Denisovans, our closely-related ancestral cousins. Those encounters are still recorded in the small percentages of Neanderthal and Denisovan DNA in the genome of many people living outside Africa. From a genetic perspective, their legacy persists today. Yet as distinct populations recognizable in the archaeological and fossil record, Neanderthals had died out by around 40,000 years ago, while some Denisovan groups lingered on longer in southeast Asia, and may have finally vanished a mere 15,000 years ago. This raises an intriguing question: can fossil remains of ancient hybrid descendants of Neanderthals, Denisovans, and modern humans be found? And if so, what happened to these hybrids?

In February 2002, explorers inside a cavern in southwest Romania known as Peștera cu Oase, or "The Cave with the Bones," squeezed through a narrow hole into a previously unknown chamber, where they came across a human jaw. It was large, with unusually big molars that got progressively bigger toward the back—characteristic Neanderthal features. But its protruding chin was typical of a modern human, an impression reinforced when most of a second skull was found in the cave the following year. Radiocarbon dating revealed the Oase bones to be around 42,000 years old, placing the Oase people among some of the earliest arrivals of modern humans in Europe.

In 2015, the plot thickened when geneticist Qiaomei Fu decoded the DNA in the Oase jaw. No less than 6–9% of it was characteristically Neanderthal, which is more than three times higher than the average of around 2 percent found in most living Europeans' genetic code today. Moreover, unusually long sequences of Neanderthal DNA were preserved intact. There had not been enough time for those genes to be sliced up and shuffled by recombination (the exchange of genes between parents that happens every time they produce offspring). Fu concluded that the DNA provided evidence of a sexual encounter between a Neanderthal and a modern human that had occurred only 4–6 generations before the lifetime of their descendant in the cave, 42,000 years ago.

But that was not all. In 2018, a new human bone fragment was identified from Denisova Cave in Siberia, the site of the original discovery of Denisovan DNA. Analysis of the genome extracted from this fragment indicated that the bone was roughly 100,000 years old and belonged to a girl of around age 13. Extraordinarily, nearly half her genes were Neanderthal and the other half were Denisovan. It meant that the girl had to be the first-generation daughter of a Neanderthal mother and a

(Opposite) The massive jaw from Peștera cu Oase Cave in Romania belonged to a 42,000-year-old male modern human. The jaw had an unusually large proportion of Neanderthal DNA, indicating a Neanderthal among the man's ancestors a mere two centuries or so previously.

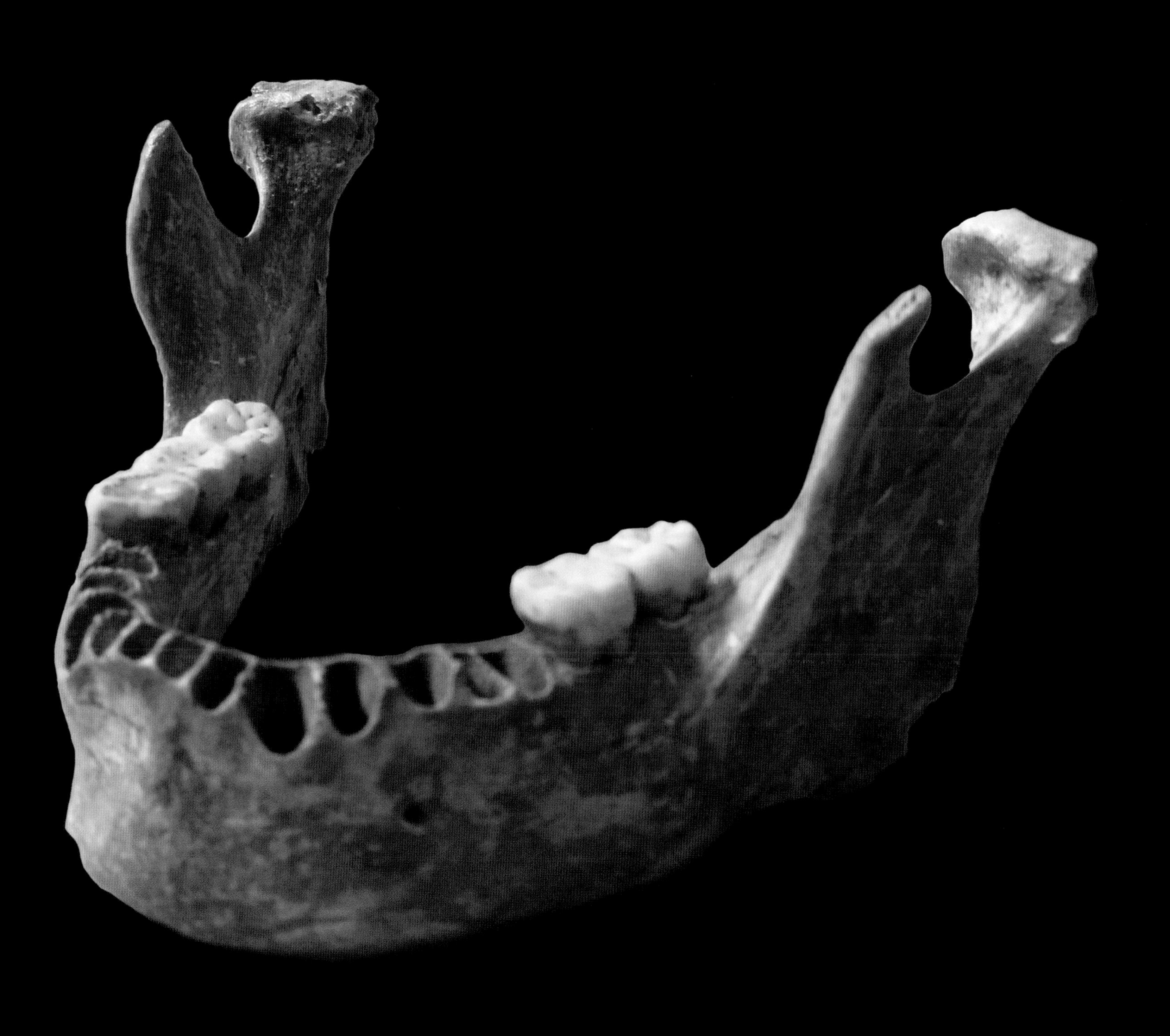

Denisovan father. Geneticist Svante Pääbo commented that it was "like finding a needle in a haystack."[110]

How widespread was such interbreeding as different ancestral groups encountered one another in Eurasia? One example is the "ghost population" recently identified in Asian genomes by an A.I. machine learning program. The analysis indicates that these mystery people originated from encounters somewhere in Asia between modern humans and hybrid Neanderthal–Denisovans. The event that produced the teenage girl in the Denisova Cave was evidently part of a wider pattern of interbreeding.

What happened to the descendants of these hybrid people? Did they flourish or rapidly die out? To take one example, the gene sequences from the Oase jaw have no special affinity to any European or Asian population today. This hybrid's lineage probably died out, to be replaced, eventually, by that of other modern humans moving into the region.

One finding offers a significant clue to the fate of the hybrids and of Neanderthal populations as a whole: their genomes were about four times less diverse than those of today's humans, almost certainly due to their small, isolated populations. A smaller, less varied gene pool meant that harmful mutations were more likely to persist and that Neanderthals probably suffered greater susceptibility to disease. We know of several cases of inbreeding, including a Neanderthal woman from Siberia whose genome had long stretches of identical paired strands of DNA. Her parents must have been very closely related, possibly even half-siblings. In the case of 13 Neanderthals whose remains were found deposited together in El Sidrón Cave in northern Spain, analysis of their mitochondrial DNA revealed that they were very closely related, most likely representing a single family. Their bones showed signs of multiple congenital disorders, which were interpreted as evidence of inbreeding. "Such inbreeding may have left Neanderthals vulnerable to high rates of genetic diseases," note molecular biologists Kelley Harris and Rasmus Nielsen, "which can threaten population health further and lead to death spirals of mutational meltdown."[111]

Meanwhile, the numbers of modern humans expanded rapidly as they spread into Eurasia. One archaeological study estimates that there were ten times more early modern human sites in Europe than Neanderthal occupations. Larger populations of modern humans could have pushed the Neanderthals into marginal living conditions and eventual extinction in some areas, while the remainder were probably rapidly assimilated by modern humans through interbreeding. "Since humans outnumbered Neanderthals by approximately ten to one during the interbreeding period," Harris and Nielsen say, "it is possible that the Neanderthals did not truly die off at all, but simply melted together with the human species. One could perhaps argue that Neanderthals did not disappear due to warfare or competition—but due to love."[112]

(Opposite) In 2004, archaeologists deep inside Romania's Peștera cu Oase Cave found some 5,000 bones of cave bears, goats, wolves, and other animals. The red dot at the left marks the findspot of a second human skull, close to the earlier discovery of the jaw shown on p. 179.

ENIGMA OF THE LION MAN

One of the world's oldest masterpieces of prehistoric sculpture, the ivory statuette known as the Lion Man is also one of the most enigmatic. Laboriously hewn from a mammoth tusk around 40,000 years ago, the one-foot-tall creature appears to be standing on its toes, its back and shoulders arched tensely as if waiting to spring into action. From the neck up, it is unmistakably a cave lion, a formidable Ice Age beast, now extinct, that was bigger than today's African lion. From the shoulders down, it is unambiguously human, complete with a navel and particularly realistic calves, ankles, and feet. According to archaeologist Jill Cook, the statuette is "powerful, mysterious and from a world beyond ordinary nature. He is the oldest known representation of a being that does not exist in physical form but symbolizes ideas about the supernatural."[113]

The survival of this statuette is little short of a miracle, as was its reconstruction from hundreds of ivory fragments over the course of half a century. The story begins on August 25, 1939, one week before the outbreak of World War II. Prehistorian Robert Wetzel had just received his military call-up papers and he and a colleague were scrambling to finish their final day of digging at Stadel Cave near the Danube River in southwest Germany. In a dark chamber at the back of the cave, they came across the ivory shards. Not knowing when, or if, they would be able to return, they backfilled their excavation with the dirt they had just dug up before hastily retreating with their find. For three decades, the pieces of the statuette lay neglected in a cigar box in the city of Ulm's museum until archaeologist Joachim Hahn began studying finds from the cave in 1969. He started the marathon process by gluing together more than 200 pieces of ivory, and, after a few days, he recognized the figure he was assembling as a hybrid of beast and human, although what kind of beast was not clear. After fresh reconstruction efforts in the 1980s, enough of the head emerged to reveal its identity as a lion. In 2009, archaeologist Claus-Joachim Kind systematically re-excavated the cave. "It was a huge surprise,"[114] he says, when he discovered nearly 600 more ivory scraps in the backfill dirt left behind by Wetzel. It took two more years to solve the immense challenge of restoring the statuette to the state we see today.

The original carvers of the figurine expended equally impressive effort. A recent experiment using replica stone tools of the period suggests that it must have taken some 400 hours to hew the figure from the tough mammoth ivory. This was clearly an extraordinary undertaking for anyone living in a hunting community, and implies that the figure embodied special social and symbolic significance. In 2002, archaeologist Nicholas Conard found a "Little Lion Man"—a similar ivory figure, but only an inch high—at the

(Opposite) Half-lion, half-human, the mysterious Lion Man statuette from Germany's Stadel Cave, carved from mammoth ivory some 40,000 years ago, may indicate a belief in shamanism—the supernatural transformation of humans into animal spirits.

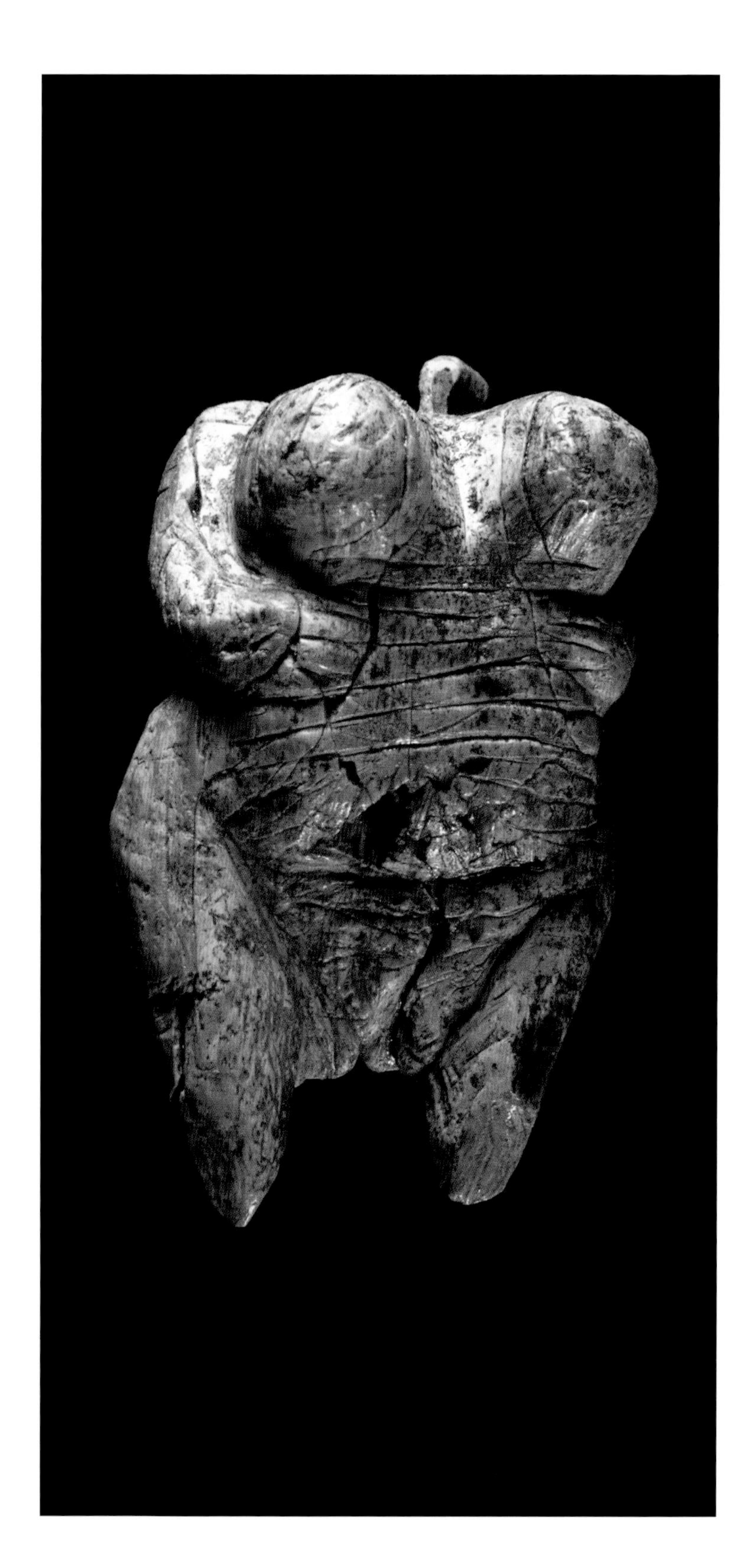

(Left) The extraordinary figurine from Hohle Fels Cave in southwest Germany, dating from around 40,000 years ago, is one of the world's oldest female images. (Opposite) A delicately crafted five-hole flute carved from a vulture wing bone was also excavated at Hohle Fels.

cave of Hohle Fels, 25 miles from Stadel. The recurrence of the Lion Man motif suggests it may be connected with a widespread myth—perhaps the transformation of humans into animal spirits that is a central preoccupation of shamanism, often considered to be humanity's most ancient strand of religious belief. Other intriguing beast-men hybrids are depicted in painted Ice Age cave art, such as a striking figure of a man-bison from Chauvet and a celebrated "birdman" figure at Lascaux. These painted images are located in dark, secluded parts of the caves; similarly, northern-facing Stadel Cave receives little light, and the recess where Lion Man was discovered contained no tools or other signs of regular occupation. This raises the possibility that all these symbols could mark the site of some type of rite or celebration.

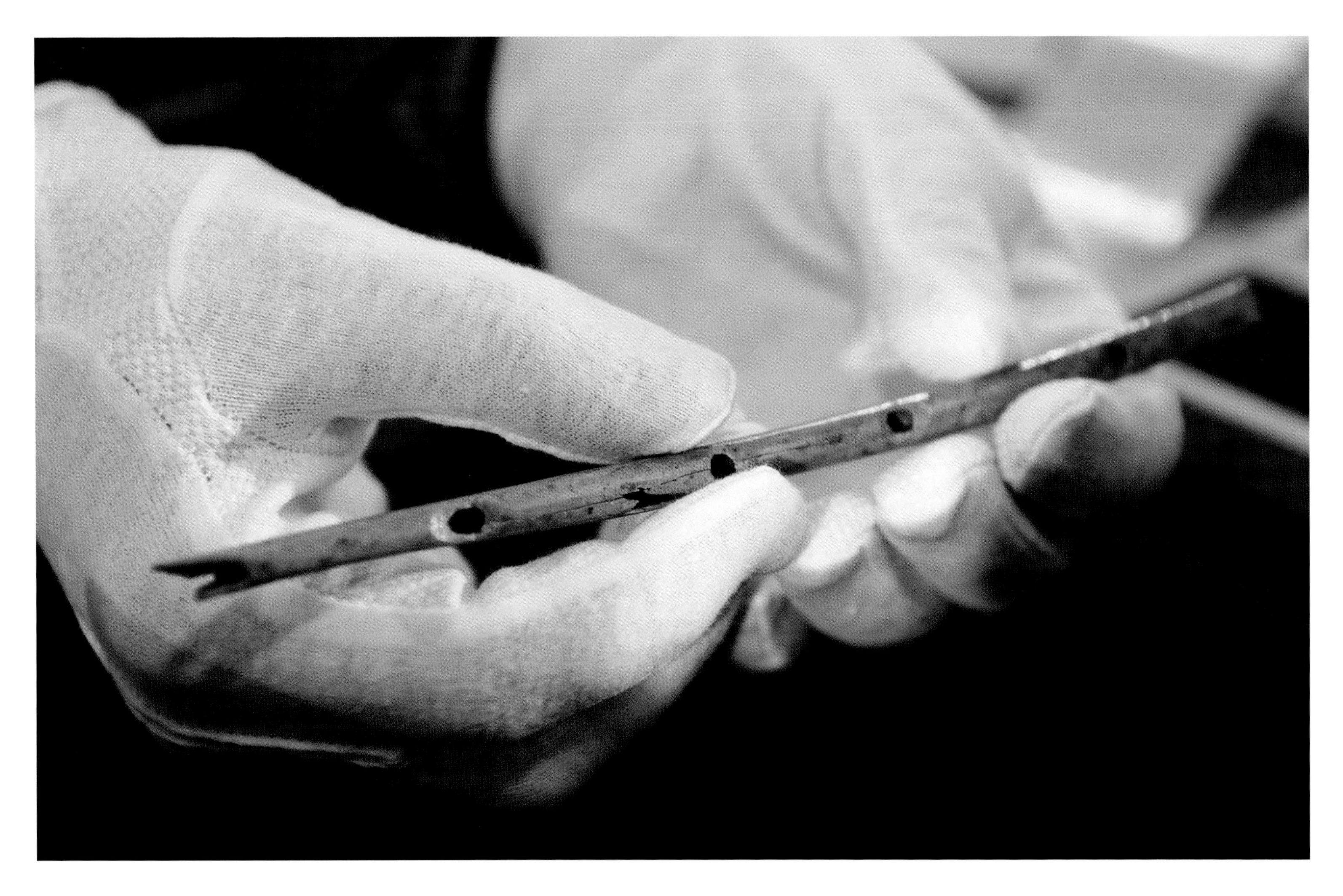

Although uniquely impressive, the Lion Man is one of many finds that testify to an extraordinary flowering of creativity among the first wave of modern humans to settle in the Danube region around 40,000 years ago. In addition to Little Lion Man, exquisite miniature ivory animals, their surfaces polished to a shine by constant handling, have been found at Hohle Fels as well as two other caves close by. In 2008, Conard's team found the so-called "Venus of Hohle Fels," the world's oldest female figurine, her ample proportions as striking as the austere pose of the Lion Man. They also unearthed a delicate flute perforated with five finger holes, the most complete of several such instruments from the local caves, all of them intricately crafted from mammoth ivory tusks or the wing bones of swans or vultures. When modern flautists play an accurate copy of the Hohle Fels flute, they produce haunting notes that suggest well-developed musical expertise. Unlike the Lion Man at Stadel, the flutes and miniature animals were found mixed in with everyday refuse rather than in an apparently special setting.

While these objects are the world's first evidence of musical instruments and figurative sculpture, they are anything but primitive. Instead, they were clearly part of a complex, fully developed world of symbolic communication. They express feelings and beliefs that are recognizable today, even if their exact meanings remain elusive.

THE BEASTS OF CHAUVET

As Jean-Marie Chauvet, Eliette Brunel Deschamps, and Christian Hillaire edged their way into the darkness, they felt a draft. They had just crawled through a narrow hole in a cliffside overlooking the Cirque d'Estre Gorge, a spectacular natural amphitheater in southeast France's Ardèche region. They were there to explore the limestone caves and overhangs that frequently sheltered humans during the last Ice Age. The air was flowing from a narrow, partly blocked, passageway. After the cave explorers had removed the obstructing rocks, Deschamps wriggled head-first down it until she saw a chamber opening up 30 feet below her. Using a rope ladder, they dropped down and found themselves in a network of chambers with breathtaking calcite formations. The floors were littered with the bones of ancient cave bears that had hibernated there.

Deschamps was the first to spot the paintings, crying out in surprise as she spotted two lines painted in red ocher, then a little mammoth. Anticipation mounting as they roamed farther, they came across a series of stunning friezes on the cave walls, depicting astonishingly realistic figures of lions, horses, and rhinos. One major mural, outlined in charcoal, features a dozen or so lions and lionesses, some of them tensely poised as if ready to spring on their prey. A few of the horses and rhinos have multiple outlines as if they were in motion, while a cluster of rhinos seem to be sparring with each other. Other figures exploit the natural contours of the rock to create an almost sculpted, 3D effect. The explorers wrote that they were "seized by a strange feeling. Everything was so beautiful, so fresh, almost too much so. Time was abolished, as if the tens of thousands of years that separated us from the producers of these paintings no longer existed."[115] Discovered by the trio of explorers on December 18, 1994, Chauvet proved to be perhaps the most impressive discovery of cave art since four French teenagers stumbled upon Lascaux in 1940.

As years of strictly controlled recording and research unfolded in Chauvet's pristine interior, cave art scholars experienced a number of shocks. The first was the choice of animals depicted. Despite existing variety in an art form that flourished for at least 25,000 years, the major decorated panels of southwestern French caves tend to feature horse, bison, and wild aurochs, all targets of Ice Age hunters. Dangerous predators are generally rare. Yet at Chauvet there are 75 powerful carnivores, such as lions, panthers, bears, and hyenas. Human figures are also absent from most caves, but in Chauvet, an overhanging rock features a strange sketch of a staring bison joined to a human body, looming over part of a female human form immediately below it. This composite image of a half-human, half-bestial form may well depict a supernatural being.

(Opposite) Archaeologist and rock art authority Jean Clottes leads efforts to conserve and analyze the masterpieces of Ice Age art discovered in Chauvet cave in France's Ardèche region in 1994.

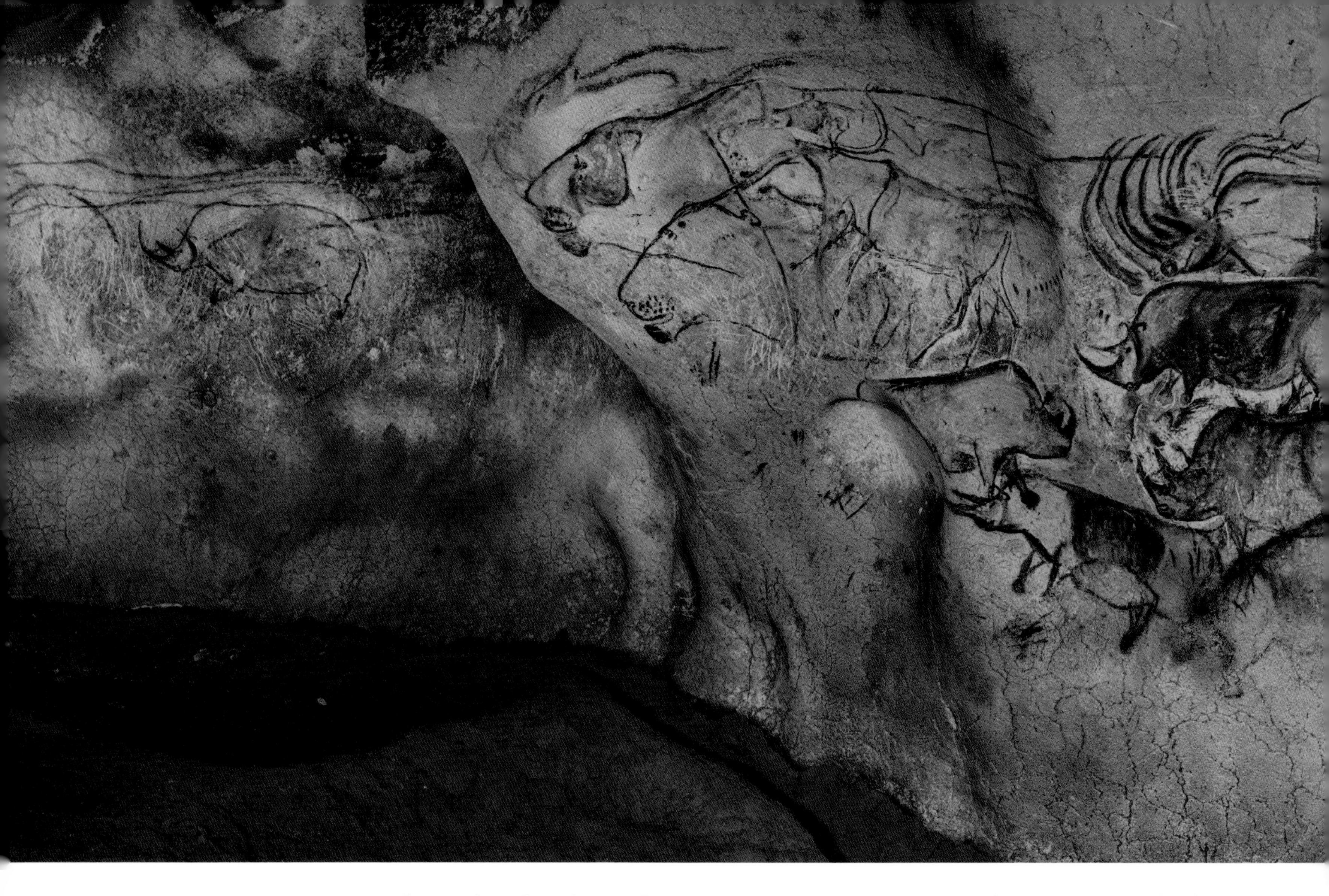

An even greater surprise came from radiocarbon dating of both the paintings and the hearths used by the artists to prepare their charcoal. The latest results indicate that most of the art is around 37,000–33,000 years old. This is some 15,000 to 18,000 years earlier than the comparably magnificent cave art at Lascaux, Niaux, and Altamira. The dates aroused widespread skepticism, for they defy long-held assumptions that the style of cave art developed gradually from simple beginnings to sophisticated masterpieces. Chauvet's dates fall during the period of the Aurignacian, the culture of the first modern humans to settle in Europe, who displaced the Neanderthals. While the Aurignacians are known for the skilled crafting of tools and personal ornaments, their cave art was thought to consist mainly of rudimentary carved animal outlines and sculpted depressions, or "cupules." With the notable exception of the extraordinary "Lion Man" and other sculptures in a few caves in southwest Germany (see p. 183), nothing in Aurignacian art approaches the summit of artistic expression so evident in Chauvet's animal friezes. Strangely, too, no evidence that Aurignacians ever lived in the region around the cave has yet been discovered.

To counter skeptics who doubted the dates, the Chauvet researchers carried out an intensive new round of tests involving multiple labs and dating techniques. Yet the results only reinforced the shockingly early age of the art. Chauvet's vivid beasts,

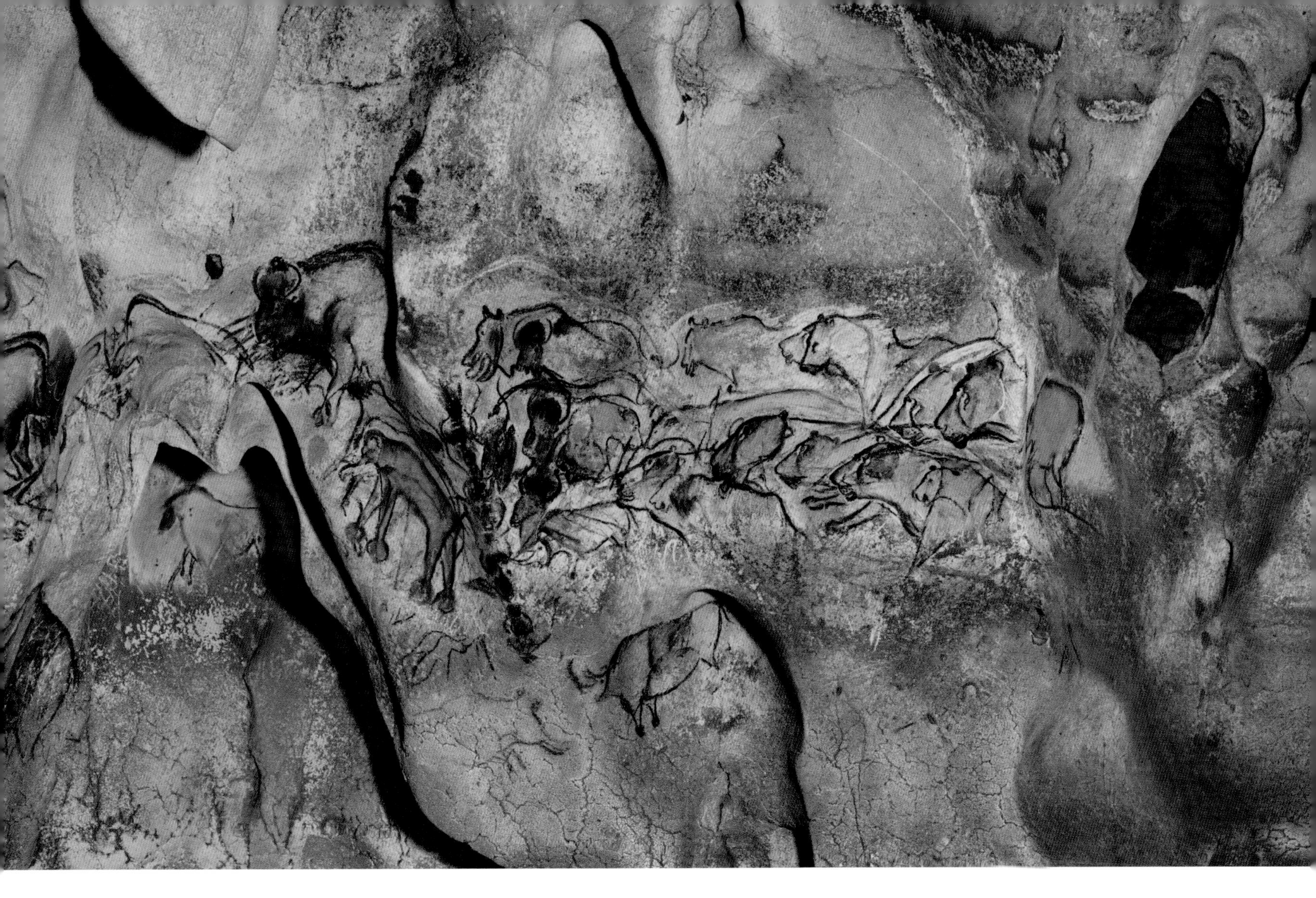

then, represent an enigma and a dilemma. Either problems remain with the dating of the site, which seems unlikely given the extent of the new tests, or common assumptions about the unfolding of artistic expertise across the Ice Age are mistaken.

The recent discovery that Neanderthals may have created cave art in Spain as early as 64,000 years ago, thousands of years before modern humans arrived in the region, further shakes confidence in old assumptions. The images in question consist of simple abstract dots and lines, and a stenciled hand (see p. 166). There is no evidence yet that Neanderthal artists depicted animals, never mind approached the skills of the Chauvet painters. Yet this new discovery serves as a warning that many of our preconceptions about the artistic skills of our ancestors may be wanting. Could other masterpieces of Ice Age art lie hidden in the soaring canyons of the Ardèche?

One of the most complex decorated panels at Chauvet includes images of bison, mammoth, and horse, together with rhinoceroses and lions rarely depicted in cave art. On first viewing this panel, "there was a burst of shouts of joy and tears," the discoverers wrote. "We felt gripped by madness and dizziness."

GRANDMOTHERS AND OTHERS

There was no "Eureka!" moment for anthropologist Kristen Hawkes, but, during the decade she spent observing the Hadza of Tanzania and other peoples in the 1980s, the evidence kept growing. "And there it was right before my eyes, you know?" she says. "Accumulating in the notebooks and day after day, these old ladies, these old ladies. I never thought they were going to be important."[116] Drawing on her observations, Hawkes and her colleagues developed an influential theory about the pivotal role of grandmothers in the human story. Their theory may help explain two major mysteries: why we live much longer than the great apes, and how the advent of modern humans triggered an "explosion" of cultural and social complexity.

Hawkes began her research at a time when hunting was viewed as the most crucial step in human evolution. It was widely assumed that our ancestors' families had thrived because successful male hunters provisioned them with meat, which was key to the growth of our big brain. But when she began to investigate the subsistence of the Hadza hunter-gatherers, she focused on the neglected contributions of women to the diet. Some 1,000 Hadza live in villages in thorny bush country, about a day's walk from Olduvai Gorge. Roughly a quarter of them are still nomadic hunter-gatherers, with no houses, crops, or cattle, and few other possessions. Women collect some of the most predictable foods on the Hadza menu, including the staple item: tubers, deeply buried starchy stems that they laboriously extract with digging sticks. To her surprise, Hawkes found that hardy postmenopausal women were the most industrious tuber diggers. Averaging seven hours a day, they were much more productive than either nursing mothers or children.

As Hawkes continued her observations, she discovered that a child's body weight was linked to its mother's success at digging up tubers. But, if a second child arrived, the amount of food their grandmother contributed was a more important factor. The more a Hadza grandmother helps to take care of dependent children, the more a family thrives. With grandma's aid, a young mother can move quickly to have another child, and these children are more likely to survive. Over many generations, grandmothers' roles as caregivers would confer a survival advantage on their families, and favor the extension of human longevity. Hawkes's colleagues devised mathematical models that helped her demonstrate the link between grandmothers' aid, shorter birth intervals, and lengthening human lifespans.

Primatologist Sarah Hrdy, renowned for her work on child-rearing and human evolution, has widened Hawkes's argument to include other potentially vital helpers, such as fathers, siblings, aunts, or uncles. In around 50 percent of primate species, mothers receive varying degrees of help from fathers or relatives but this sharing of infant care rarely extends to actual nursing or provisioning of others' young. In fact, our practice of cooperative child rearing stands in stark contrast to that of most mammals. So why and how did the nearly universal human pattern of cooperative care develop?

Human mothers can have many more children during their fertile years compared to great apes, who can only raise one infant at a time. (Today's hunter-gatherers average 3–4 years between births while, for great apes, the interval is about six years.) Hrdy proposes that cooperative child rearing is the only explanation for how our slow-growing, big-brained, and highly dependent offspring could possibly have developed in the first place. As Hrdy says, "An ape that produced such costly, slow-maturing offspring as we could not have evolved unless mothers had a lot of help."[117] The dependency of human infants on other

family helpers besides mothers also explains why babies are so tuned in to reading faces and social cues such as smiles or frowns at a very early age. “Long before they can speak,” notes Hrdy, “human infants obsessively monitor intentions and are eager to learn what someone else thinks and feels...Youngsters just a little better at monitoring the mental states of others, at appealing to and soliciting nurture from them, would be the best cared for, best fed, and most likely to survive.”[118]

The impact of cooperative caregiving would have extended far beyond the realm of raising children. The increasing survival of elders—grandfathers, too—must have contributed to the transmission of culture, since they would help pass on vital knowledge such as the location of food or the risks of toxic plants, as well as skills such as tool-making. By turning the spotlight on the overlooked role of women and the elderly, Hawkes's influential theory opened up a bold new perspective on the processes that made us fully human.

Since the 1980s, the nomadic Hadza hunter-gatherers of Tanzania have been a focus for debates about the significance of grandparents in human evolution. Here, a young woman hands her grandmother her newborn infant.

BABOONS' PERILOUS CHILDHOOD

Every morning, groups of baboons climb down from their sleeping trees to begin their day on the Kenyan savanna. Overlooked by the majestic peak of Mt Kilimanjaro, which rises to the south across the border in Tanzania, the grasslands of Amboseli National Park are a vast and arid expanse, with a five-month-long dry season and highly unpredictable conditions the rest of the year. Six days a week, fifty-two weeks a year, the Amboseli baboons are shadowed by a team of dedicated Kenyan researchers who have kept continuous records of daily behavior for the last 48 years. Comprising comprehensive histories of more than 1,500 animals, the Amboseli Baboon Research Project is one of the most detailed studies of any wild primate. This unique trove of data is allowing researchers to answer questions with resonance for humans as well as baboons: how does social status affect health and longevity? What impact do traumatic childhood events have on mothers and the nurturing of the next generation? And can strong social bonds help overcome the negative effects of early hardship?

Four decades of research have shown that the Amboseli baboons are highly intelligent and complex creatures, acutely tuned to the social forces within their communities but also vulnerable to many challenges to survival—particularly in infancy.

(Opposite) A mother and infant baboon at the Amboseli Baboon Research Center in southern Kenya, where insights into the long-lasting effects of childhood trauma and deprivation have been gained across multiple generations of baboons.

Infant baboons must learn to forage for more than 250 types of leaves, roots, fruits, and seeds, avoid lethal diseases and predators, and navigate the intricate social map around them. During periods of drought, first-year infant baboons have only a 50/50 chance of making it. In a landmark study, the Amboseli team investigated how infants fared in response to half a dozen different threats, including losing their mother, drought, competition from siblings, and rising population pressures. If infant baboons were lucky enough to experience only one of these setbacks—the team nicknamed them "silver spoon kids"—they would typically survive into their late teens or early twenties. But if they had three or more strikes against them, they often died by age 9, their life expectancy cut in half—"a shockingly large effect," says project co-leader Susan Alberts, "for an animal that lives about 27 years at most."[119]

The advantages of a "silver spoon" upbringing are evident in the story of Kelly, born into a middle-ranking but strongly bonded family. She prospered from two years of undivided attention from her mother, Kupima, before her brother Ken was born. Both her mother and other relatives were around her until she reached adulthood, and solidified their relationships by frequent grooming. Kelly lived to be over 26 years old, one of the

DUKE
GRADUATE SCHOOL

(Opposite) Three generations of primatologists have made daily observations of baboon activity at Amboseli for nearly half a century, notably (left to right) Elizabeth Archie, Susan Alberts and Jeanne Altmann.

oldest females in the entire study. In stark contrast, two females named Puma and Mystery were both born during drought years. Their mothers were low-ranking and died before their third birthdays. Puma fell victim to a leopard attack at age 7, while Mystery disappeared at age 14, presumably also the victim of a predator. Mystery's orphaned infant died soon afterwards.

A striking result of the Project's latest work, supported by The Leakey Foundation, is that early hardships inflicted on baboon mothers carry over into the next generation. Alberts explains that "if females experience early adversity themselves, then their offspring—even if they *don't* have the same kind of hardships—have compromised survival during their younger years."[120] The team is exploring possible explanations for this cross-generational effect, such as the idea that a mother's tough childhood might result in lower quality care given to her infants.

Another possibility is that genetic mechanisms could be at work. In human populations, babies born during periods of famine have higher than normal rates of health problems in later life. For instance, during the final period of German occupation of the Netherlands near the end of World War II, severe food shortages led to widespread starvation and about 18,000 deaths. Infants born during the famine suffered unusually high rates of diabetes, heart disease, obesity, and other ailments. Studies suggest that changes in the way genes are regulated and expressed, triggered by the famine and known as epigenetic effects, could have been responsible.

A similar possibility is that drought or social stress inflicted on baboon mothers could lead to the epigenetic "silencing" of certain genes, adversely affecting the health and survival of their offspring. An innovative study by Leakey grantee Amanda Lea provides strong evidence that shifts in baboon diet *do* result in changes in gene expression. Her work opens the door for future investigations of the biological mechanisms that may help explain the connection between low social status and poor health.

Can childhood adversity be overcome? "If you're off to a bad start in life," says Alberts, "whether or not you can change your trajectory is a burning question for both human and wild animal studies."[121] The Amboseli Project has already demonstrated that strong social connections are associated with longer lifespans. Now, the team is investigating exactly how the nurturing and protection that a young animal receives from parents, relatives, and the wider social group can offset early disadvantages. As Alberts writes, "The saying, in human societies, that 'it takes a village' to raise a child turns out to have strong resonance throughout our primate family tree: social ties can determine whether a young animal lives or dies just as surely as an outbreak of disease or a lurking predator. Our primate cousins, like ourselves, are essentially social creatures in the deepest sense of the word."[122]

INSIDE BABOON MINDS

Royal, a grumpy old male, is one of an 80-strong troop of baboons in the Moremi Game Reserve in Botswana. Like all baboons, he belongs to a complex, highly competitive society, and pays constant attention to the social cues around him. His world, much like ours, depends on communication, and understanding its rules sheds light on our own social evolution. Today has been uneventful until, somewhere off in the bushes to his right, Royal hears the unmistakable whoop that female baboons make after mating. Her call is of little interest because Royal knows it belongs to Jackalberry, the partner of Cassius, a high-ranking male who guards her from other suitors. But then there is a surprise. To Royal's left, far away from Jackalberry, comes the familiar grunt of Cassius. Royal seems to ponder the implications for a moment: could Jackalberry be cheating on Cassius with somebody else? Perhaps hoping for some action himself, he heads off in her direction. Sadly for Royal, instead of a receptive female hidden in the bushes, he finds the true source of Jackalberry's call: a loudspeaker planted by humans.

Why would researchers play a trick on a crusty old baboon? Dorothy Cheney and Robert Seyfarth, a husband-and-wife team of primatologists, spent 16 years at the Moremi Reserve living in tents, occasionally being chased into trees by lions, and devising ingenious experiments that reveal how baboon minds work. Their society is rigidly organized around matrilineal families, and females stay throughout their lives in the group where they were born. Social status is inherited and long-lasting, but through grooming favors and sheer force of personality, a low-ranking female can make alliances and friendships with her superiors.

(Opposite) Baboons resting and grooming in Botswana. Dorothy Cheney called them "voyeurs, essentially...you can show that the animals recognize not only their own relationships and dominance ranks, but also those of others."

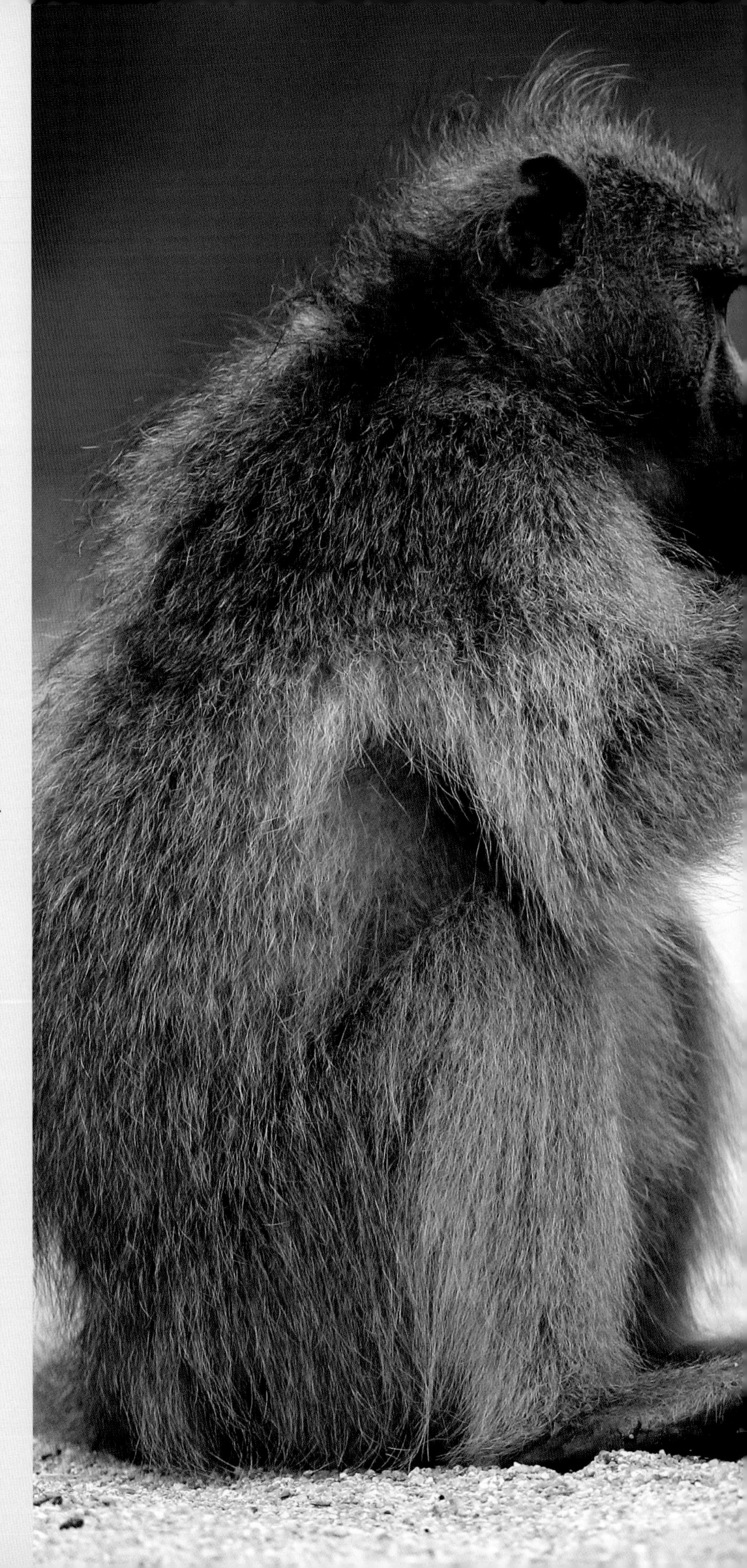

These ties offer protection from predators and aggressive males, boosting survival and reproductive success. Cheney and Seyfarth wrote that "monkey society is governed by the same two general rules that governed the behavior of women in so many 19th-century novels. Stay loyal to your relatives...but also try to ingratiate yourself with the members of high-ranking families."[123]

Cheney and Seyfarth spent months acquiring a library of baboon calls. By playing their recordings back from hidden loudspeakers, they discovered that an individual recognizes the voices of all 80 members of the troop. Then they embarked on experiments that violate the normal rules of baboon communication. For example, the social wheels of baboon society are greased by a strict protocol of how they address one another. When a high-ranking animal encounters a lower-ranked one, it sometimes issues a standard "threat grunt;" the lower-ranked animal typically responds with a deferential scream. With the help of their recording library, the researchers would scramble the order of these calls or combine them in unusual ways, then measure the surprised reactions of baboon listeners.

Years of similar "playback" experiments gave Cheney and Seyfarth extensive data on the social knowledge and expectations of each animal. They concluded that baboons understand not only their own place in the hierarchy but the status and relationships of everyone else, too. As the incident with Royal revealed, they

(Opposite) Biologist Dorothy Cheney at the Moremi Game Reserve in Botswana, where she and her husband Robert Seyfarth carried out ingenious experiments to test the awareness of kinship and rank among baboons.

are keenly interested in transient day-to-day shifts in relationships in a way irresistibly reminiscent of our own obsessions with gossip on Facebook and Twitter. We, too, are highly social primates. Cheney and Seyfarth argue that the challenge of life in a social group shaped the evolution of our minds no less than it did baboons'.

Yet human and baboon minds are crucially different. Despite deep understanding of their social network, baboons seem lacking in empathy; they apparently do not grasp what other animals know or how they are feeling, which psychologists refer to as "theory of mind." According to Cheney and Seyfarth, "Baboons' theory of mind might best be described as a vague intuition about other animals' intentions."[124] They observed many striking examples of this deficiency. In one incident, flooding drove the adults in the troop to swim to another island, leaving nearly all the young animals stranded behind them. While the juveniles emitted distress calls that were clearly heard by the adults, only one adult ever responded. For three days, the vulnerable juveniles banded together to survive, finally braving the water to swim across and reunite with the group. In this case and many others, baboons show little interest in the mental states and needs of others, even their own infants.

Another contrast can be found in the limitations of their vocal output. The playback experiments clearly demonstrate that they can recognize the order of two sounds and attach meaning to it, almost like a sentence. Despite this basic grasp of syntax, their calls consist only of single sounds that are never combined to express new meaning—the essence of human language. "The ability to think in sentences does not lead them to speak in sentences," Cheney and Seyfarth noted.[125] "Under natural conditions they certainly have interesting vocalizations. But if they have such a rich understanding of other animals' calls, why do they produce so few? It's a real puzzle, because the same animal is both a creative listener and a very limited speaker. Where does this limitation come from? The speech apparatus? The brain? Or is it just a matter of motivation?" After years of dedicated and inventive research, Cheney and Seyfarth had no clear answers to the riddle. Yet their painstaking experiments suggested that some of the foundations of language, one of humanity's most significant evolutionary advances, were already laid down among the common ancestors of baboons and humans, dating back many millions of years in our shared past.

THE HOBBIT ENIGMA

The island of Flores is the biggest in a chain of rugged volcanic islands strewn across the world's deepest seas, halfway between mainland Asia and Australia. Some 100,000 years ago, it was home to some very odd creatures: *Stegodon*, an extinct species of pygmy elephant about the size of a buffalo; giant rats as big as rabbits; the fearsome Komodo dragon, today's largest land reptile; a six-foot-high carnivorous stork; and *Homo floresiensis*, a tiny ancient human around three-and-a-half-feet tall, whose discovery in 2003 created an immediate sensation. Nothing like its peculiar features had ever been seen in the fossil record—not least, its huge feet. Three-quarters the length of its thighbone, they immediately suggested its nickname, the Hobbit, inspired by the character in J. R. R. Tolkien's *The Lord of the Rings*. What was it doing here on a remote Indonesian island together with such a bizarre bestiary of other extinct species?

The Hobbit's tale begins in the 1990s, when Mike Morwood, an archaeologist based in Australia, decided to look on Indonesian islands for traces of early modern humans who first arrived in Australia at least 60,000 years ago. On Flores, he first visited a site called Mata Menge, where a Dutch archaeologist in the 1960s had reported 750,000-year-old stone tools alongside *Stegodon* bones, a claim that few had taken seriously. Morwood's team found more tools, confirmed the early

The entrance to Liang Bua (the "Cool Cave") on the Indonesian island of Flores, where the 2003 discovery of the "Hobbit" created a sensation and posed one of paleoanthropology's most perplexing riddles.

(Opposite) A modern human skull (left) next to that of LB1, the most complete of the "Hobbits" found at Liang Bua. At just 380 cc, the Hobbit's brain was smaller than the average chimpanzee's and less than a third the size of our own. Yet a cast of the skull's interior indicated that the brain's organization was surprisingly advanced.

date, and then moved on to Liang Bua—the "Cool Cave"—a cathedral-like limestone cavern. Here, just as the dig was due to close in 2003, local excavator Benyamin Tarus unearthed the complete skull, pelvis, both upper and lower limb bones, and parts of the hands and oversized feet of a tiny human, known as LB1. Its fragile bones had the consistency of wet paper. The femur was slightly shorter than that of Lucy the australopithecine, indicating a height of just 42 inches. Eventually, the team recovered a second jaw and isolated teeth and bones belonging to at least 14 other individuals. It was at first estimated that the Hobbit had survived to the astonishingly recent date of 12,000 years ago.

LB1's most controversial feature was its miniscule brain. At around 400 cc, it was only a little over twice the volume of a tennis ball, around the size of a chimp's brain. This startling fact launched a bitter academic debate: was *Homo floresiensis* a valid ancient species or simply an anatomically modern human suffering from a pathological condition? The skeptics drew attention to various abnormalities that result in small brains, such as microcephaly, congenital hypothyroidism, and Laron syndrome. Others rebutted the argument by pointing to the shape of the brain as revealed by casts of the skull's interior. From this evidence, it seems that the brain was arranged in a surprisingly advanced manner, including a well-developed prefrontal cortex normally associated with high-level thinking and forward planning. That was perhaps not so surprising, since the Liang Bua dig showed that the Hobbit was a competent stone toolmaker, and was smart enough to have hunted down intimidating beasts such as *Stegodon* and Komodo dragons for its supper.

In 2016, two important new discoveries finally settled this acrimonious debate. A major reinterpretation of the soil layers at Liang Bua led to new dates for LB1, now thought to have occupied the cave between 100,000 and 60,000 years ago. While the timing of *Homo sapiens*' arrival on Flores is still unknown, the idea that the Hobbit was a deformed modern human seems even more unlikely. The second discovery emerged from renewed excavations at Mata Menge, where new human remains were found: part of an adult jaw and milk teeth from two different children. According to some reports, the Mata Menge jaw is even smaller than LB1's, although it is too fragmentary to be sure if it really does belong to the Hobbit's species. Nevertheless, it looks like more traces of the diminutive hunter and toolmaker remain to be found in Flores's spectacular volcanic landscape.

The peculiarity of *Homo floresiensis* and the other creatures that flourished on Flores can partly be explained by their isolation. Even during periods of low sea levels, the island was never connected to mainland Asia or Australia; reaching it involved crossing treacherous currents and deep channels, one at least 15 miles wide. Only birds and bats could easily cross that gap; other species may have rafted across on vegetation in a fluke event such as a tsunami. Did the Hobbit reach Flores accidentally or on a deliberately constructed raft? Until its discovery, it was widely assumed that only modern humans had the skills and intellect to plan an ocean voyage. In any case, once on Flores, the new

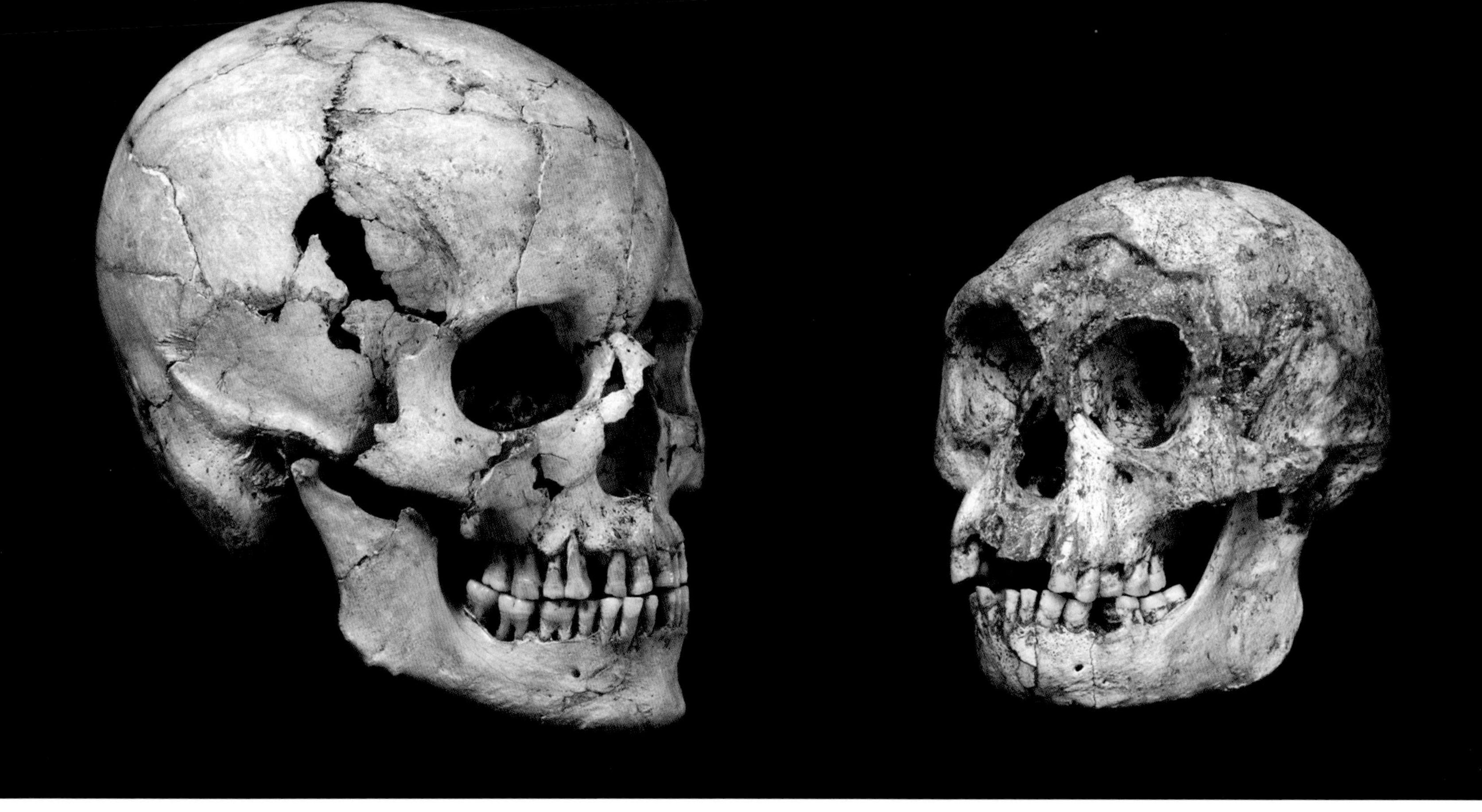

arrivals were subject to the special evolutionary pressures of remote islands, which tend to favor oversized versions of certain species and miniature forms of others. No one can be sure if the Hobbit was already tiny when it first arrived, or whether it progressively shrank due to the "island dwarfing" effect that produced the pygmy elephant *Stegodon*.

So where did the Hobbit come from? Close to Lucy's size and sharing some of her ancient tree-climbing adaptations, LB1 hints at the possibility of a previously unknown exodus from Africa perhaps 2 million years ago, long before the mastery of fire and the advent of big brains once thought to be essential for our ancestors to leave their African cradle. A contrary theory proposes that they *were*, in fact, descended from the later big-brained *Homo erectus*, but owe their distinctive features to the pressures of isolation and island dwarfing.

Whatever the case, new evidence suggests that the Hobbits were not a solitary, one-off evolutionary experiment. Since 2007, fossil remains of another tiny human have been unearthed in Callao Cave on the island of Luzon in the Philippines, nearly 2,000 miles from Flores. The discoverers argue that it represents a new species, *Homo luzonensis,* although so far the evidence consists of just 13 fragmentary fossils, including hand and foot bones and part of a femur from three different individuals. Like the Hobbit, the hand and foot bones show some of the tree-climbing adaptations of Lucy and other australopiths. Dating to at least 50,000 years, the Callao cave dwellers may turn out to be an offshoot of the same population as *Homo floresiensis,* colonizing Asia's Pacific islands at an astonishingly early date. This recent discovery suggests that other traces of an unknown migration, spanning vast distances across southern Asia, are likely to be found. The region's relatively untouched fossil wealth may harbor many more surprises.

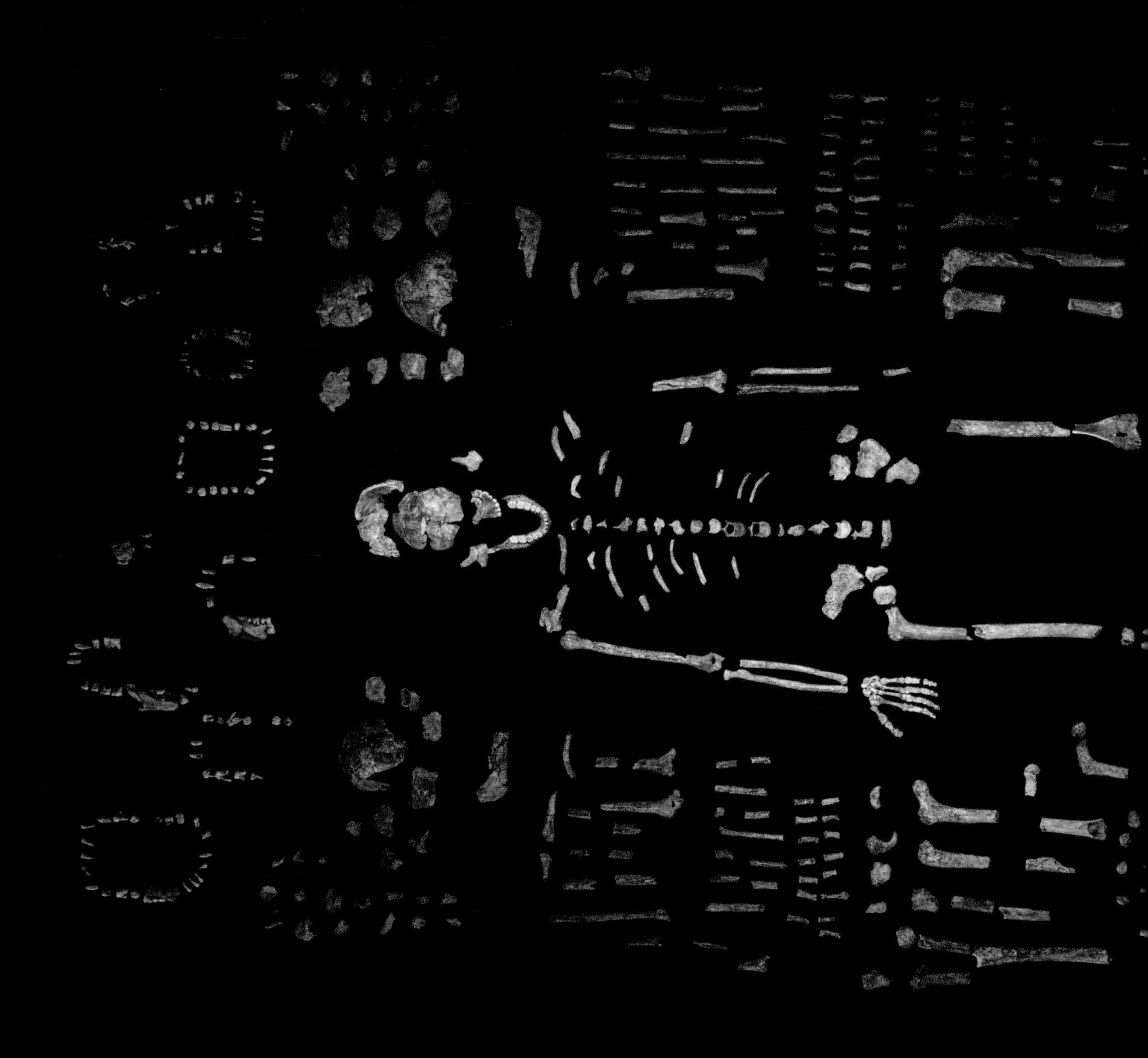

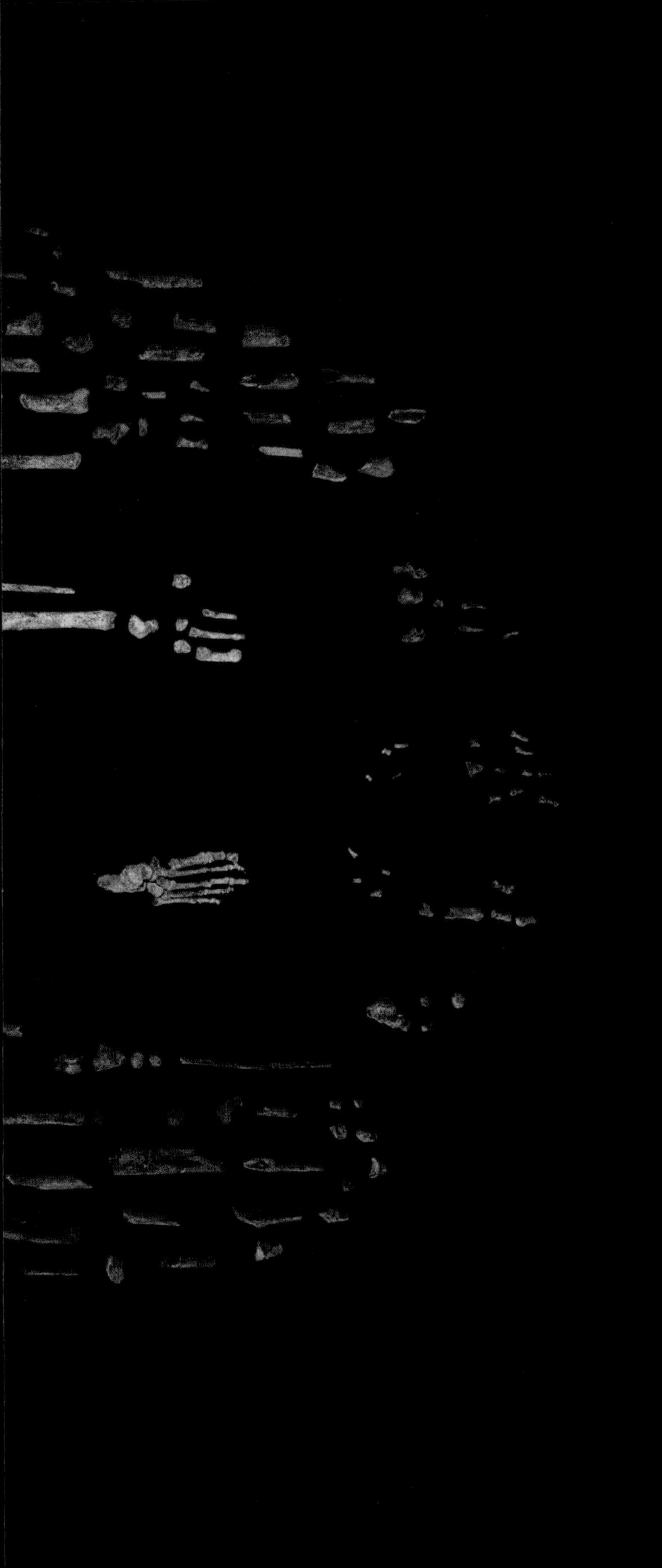

THE CHAMBER OF STARS MYSTERY

The Facebook post, describing an urgent need for expert volunteers to join an unspecified project in South Africa, appeared in October 2013. There was just one catch: "The person must be skinny and preferably small. They must not be claustrophobic, they must be fit, they should have some caving experience, climbing experience would be a bonus."[126] Alia Gurtov, a young American graduate student and Leakey grantee at the University of Wisconsin, was intrigued. She responded immediately to the author of the post, paleoanthropologist Lee Berger at the University of Witwatersrand. A few weeks later, Gurtov became one of six "underground astronauts," inching and crawling her way to a spectacular new fossil discovery deep inside Rising Star Cave near Johannesburg.

They faced daunting obstacles on their way down. First, they had to squeeze on their bellies through the "Superman Crawl," then skirt the sheer drop-offs of the "Dragon's Back," and finally squirm feet-first down "The Chute," a 40-foot vertical crack, at one point only 7½ inches wide, where Gurtov says she was "splayed out like a flying squirrel." The Chute ended with a ten-foot drop to the floor, leading to a chamber that Berger named Dinaledi, or "Chamber of Stars" in the local Sesotho language. Gurtov recalls glimpsing the chamber floor, "and there are just

(Opposite) A composite skeleton of *Homo naledi*, a previously unknown human relative, is surrounded by some of the hundreds of other fossil remains found in the almost inaccessible Dinaledi chamber at Rising Star Cave, South Africa, in 2013.

(Opposite) Paleoanthropologist Lee Berger's daughter Megan and Rick Hunter, one of the original discoverers of the Rising Star fossils, squeeze their way down the narrow chute leading to the Dinaledi chamber.

bones everywhere. There are teeth, glinting like little pearls in the dirt and identifiable bones just lying all over the place."[127]

Two "astronaut" excavators at a time took shifts in the narrow chamber, linked by a continuous video feed to Berger and his support team on the surface. They limited their investigation to a small portion of the floor, less than three feet square and eight inches deep. It was densely packed with fossils. By the end of three weeks, they had recovered a staggering total of more than 1,200 bones, later found to represent more than 15 individuals, including nine juveniles and four or five infants. Although they had dug only a tiny fraction of the Dinaledi chamber, Berger's team had recovered one of the largest and most complete samples of any fossil human relative.

The team named the new species *Homo naledi* and it was an intriguing mix of primitive and advanced features. Males averaged around five feet and weighed around 100 pounds. The pelvis shows that *naledi* was bipedal and its lower leg and foot were almost completely modern. Yet, like the australopithecines, its ape-like shoulder blades and long, curving fingers were well-suited for tree climbing. And while its teeth and jaws were much smaller than Lucy's, its braincase was not much bigger; at 500–600 cc, it was under half the modern average, and its low, sloping forehead looked nothing like our own.

With its unique blend of features, *naledi*'s place in the human story stirred conjecture: could it be a direct early ancestor of *Homo*, perhaps a million or two years old? Or an isolated late survivor on a dead-end branch of our tree, like the equally puzzling Indonesian Hobbit? It took more than two years to pin down its age until finally the analysis of three *naledi* teeth indicated a stunningly recent date: a mere 235,000–335,000 years old. By this time, the first modern humans were already emerging in Africa and their small-brained ancestors were thought to have been long extinct. The dating made the survival of its primitive features all the more striking.

But an even bigger puzzle was the mystery of how so many bodies ended up in the almost inaccessible Chamber of Stars. Some have argued that an easier, long-lost passage through the cave must have existed, but a meticulous geological study has failed to find any such evidence. The chamber was not a carnivore den, since the bones bear no gnaw marks that might have been left by predators. Nor do the bones show signs of trauma, so *naledi* was not a victim of a prehistoric massacre. Parts of a leg, hands, and feet were still in anatomical order, indicating that at least some bodies had decayed in place and had not been washed in from some other location.

Lacking any better explanation, the team was driven to a provocative and controversial conclusion seemingly at odds with *naledi's* orange-sized brain. "To date," Alia Gurtov says, "our best supported hypotheses all involve a kind of behavior that we generally attribute to humans, which is intentional treatment of the dead."[128] Improbable as it might seem, the team argues that *naledi* dragged its dead through these tortuous passages to the remote chamber as an intentional and repeated act. In 2017, the discovery of more bodies directly underneath the Chute appeared to strengthen the case that they had been deliberately brought there.

Rising Star remains a singularly enigmatic find, but one thing is clear: like many other recent discoveries, *naledi* has confounded expectations that the saga of human evolution was a straight-ahead march toward bigger brains. "It was always just a tale," says Lee Berger, "and it's ended now."[129]

THE ROOTS OF WAR

The victims' remains were obvious, scattered all over the dark desert gravel. "We could see coming out of the ground the broken lower legs, the shin bones of somebody," anthropologist Marta Mirazón Lahr recalls in a NOVA interview.[130] The next thing she saw was "the wrist sticking vertically out of the ground. So we started clearing and excavating and the bones of another one appeared, so we would clean that one, and the bones of yet another would appear. We found the remains of 27 people." Strewn over an area twice the size of a Manhattan city block, the bones were not buried but lay where they had fallen along an ancient shore line of Kenya's Lake Turkana, at a place today called Nataruk. They came from at least eight males and eight females, along with fragments of six children. A dozen adults were relatively intact and, of these, four appeared to have had their hands bound, including a woman who was carrying a nine-month-old fetus. Ten had multiple injuries, including blunt-force trauma to the head, broken hands, knees and ribs, and arrow wounds and points lodged in skulls. The arrow points were crafted from obsidian, a stone rarely found in the area, suggesting the attackers may have come from a distant community. "The injuries suffered by the people of Nataruk—men and women, pregnant or not, young and old—shock for their mercilessness," Lahr says.[131]

The 10,000-year-old Nataruk bones are also notable as the earliest firmly dated evidence of warfare. They are from a time well before the emergence of farming and settled, stratified communities, developments long assumed to have given rise to inter-group conflict. Is war a relatively recent phenomenon, born of complex societies, or are its roots hard-wired in our evolutionary past? Although the Nataruk people were hunters and foragers, the rich resources of the fertile lakeshore were evidently worth fighting for, Lahr believes, "whether it was water, dried meat or fish, gathered nuts or indeed women and children. This shows that two of the conditions associated with warfare among settled societies—control of territory and resources—were probably the same for these hunter-gatherers, and that we have underestimated their role in prehistory."[132]

The site is a key exhibit in the long-running debate about the origins of warfare. Primatologist Richard Wrangham, renowned for his long-running chimpanzee studies in Uganda, argues in his popular book *Demonic Males* that we are "the dazed survivors of a continuous five-million-year habit of lethal aggression."[133] Wrangham bases his view on the closeness of our ancestry with that of chimps, and the fact that they, like us, are among the only mammals that attack groups of their

(Opposite) Fossil hunter Justus Edung and archaeologist Marta Mirazón Lahr excavate the remains of one of the dozen men, women, and children brutally slain 10,000 years ago at the Nataruk site near Lake Turkana, Kenya.

(Opposite) The skeleton of a male victim, lying face down in the sands of the former lagoon at Nataruk. The man had severe head injuries inflicted by a blunt instrument that probably also broke his neck.

own species. Natural selection, he proposes, favored male violence deep in our shared past. As analysis of the "Great War" at Gombe has confirmed (see p. 63), chimpanzee raids on neighboring groups are often driven by power struggles between dominant males, enabling the aggressors to broaden their territory and acquire new females and fruit trees. In a similar process, according to Wrangham, the violent behavior of our male ancestors enhanced their reproductive success.

This biological view of the roots of human violence is shared by many scholars, notably Stephen Pinker in his influential book *The Better Angels of Our Nature*, which argues that the growth of civilized values and institutions since the Enlightenment has actually led to increasingly peaceful conditions in much of the globe. Pinker underpins his case by contrasting recent rates of war with the "anarchy" of prehistoric societies, in which "chronic raiding and feuding characterized life in a state of nature."[134] Recently, Pinker has drawn attention to a study of over 600 human populations from all periods, including excavated archaeological sites with human remains. Its conclusion estimates that the risk of violent death in the prehistoric past was, on average, up to three times greater than a widely cited figure for our present-day world. Such estimates are inevitably controversial, given the patchiness of the archaeological and fossil record. Cultural and environmental forces, say Pinker's critics, are more significant drivers of human warfare than any innate aggressive urges.

Yet homicidal behavior unquestionably dates back long before the Nataruk massacre. The oldest known case of interpersonal violence comes from the Sima de los Huesos (or "Pit of the Bones") in northern Spain (see p. 162), where at least three of the 430,000-year-old skulls show the marks of deliberately inflicted trauma, including one penetrated twice by the same weapon. Among the Neanderthals, a healed fracture in the St. Césaire skull from southwest France is consistent with a close-range attack by a stone weapon. However, similar marks of lethal force are apparent on only one other Neanderthal skeleton among the hundreds recovered from Europe, Asia, and the Middle East. Although cannibalism has been well-documented at caves in southern France, northern Spain, and Croatia, there is no way to be sure if the cutting-up of human flesh was a hostile act inflicted on a vanquished enemy, part of a burial rite, or simply due to the demands of starvation.

Many have turned to ethnographic accounts of hunter-gatherers as comparisons for prehistoric behavior. Given that such societies are not "living fossils" of the past, and are rarely isolated from the modern world, their relevance to the debate is equally contentious. Yet the record of violence among some mobile foragers, notably Aboriginal Australians, hardly encourages the belief that our pre-agricultural ancestors enjoyed an idyllic, conflict-free existence. Eyewitness accounts by an English observer, William Buckley, in 1803, only fifteen years after indigenous Australians were first contacted, include harrowing chronicles of raids, massacres, and large-scale battles involving multiple tribal groups. Buckley recounts how one night-time surprise raid targeted a sleeping group camped at a lakeside. Next morning, "on our arrival," Buckley says, "a horrid scene presented itself, many women and children laying about in all directions, wounded and sadly mutilated."[135] Afraid the attackers would soon return, relatives hastily left the dead where they lay. Buckley's tale eerily echoes what may have befallen the victims at Nataruk thousands of years ago.

SKIN DEEP: THE ENIGMA OF SKIN COLOR

"Within our single, recently evolved species," observes biological anthropologist Nina Jablonski, "skin colors make up an exquisite palette, varying in almost imperceptible degrees from the palest ivories to the darkest browns."[136] Until recently, scientists could only speculate about how we acquired our diversity of skin tones. Apart from exceptional cases such as Ötzi, the famous "Iceman" found in the Italian Alps, or the "bog people" of northern Europe's wetlands, human skin rarely survives from the ancient world, and these cases are at most a few thousand years old. But in the last decade, DNA evidence has brought theories about how we evolved our unique "sepia rainbow" into much sharper focus.

In common with all primates today, our early ancestors almost certainly had light skin underneath a protective covering of fur. At some point, once we had left the forest and moved out onto the plains of East Africa, we became more physically active, lost our fur, and began to control our body temperature more efficiently by sweating, which enabled us to extend the range and duration of our hunting and scavenging. But without fur, our skin was fully exposed to the sun's harmful effects. It is well-established that dark skin offers protection against these effects due to the increased presence of a chemical protein called melanin. When the sun's rays stimulate its production, it collects in little pockets or melanosomes, which shield the DNA and other vulnerable biomolecules inside cells from ultraviolet damage. (Melanosomes also migrate from the base of hair follicles into the hair shaft to give each of us our distinctive shade of hair color.) We had to evolve dark colors to protect our exposed skin from skin cancer—or so the theory went. However, skin cancer normally strikes later on in life, well after women stop reproducing, so the explanation makes no evolutionary sense.

A better answer dawned on Jablonski while she attended a lecture on birth defects. She learned that some of these, like spina bifida, are linked to shortages in the body of folate, a form of vitamin B9 essential for synthesizing DNA. She also recalled reading that the sun's harmful ultraviolet rays can destroy folate in the tiny vessels that supply blood to our skin. Putting the two ideas together, it was suddenly obvious to her that natural selection would favor darker, more protective skin in regions nearer the equator, where sunlight is most intense. "I sat there literally bouncing around in my chair," she recalls, "this is a connection between pigmentation and reproductive success."[137]

In Jablonski's theory, dark skin was favored as a protection against birth defects. Then, as humans migrated out of Africa

(Opposite) Some of the 4,000 portraits of volunteers forming *Humanae*, a "colossal global mosaic" by artist Angélica Dass. She matches the skin tones of the portraits to similar shades in the Pantone catalog to reinforce the riotous diversity of human skin colors. "Using this scale, I am sure that nobody is 'black,' and absolutely nobody is 'white,'" says Dass.

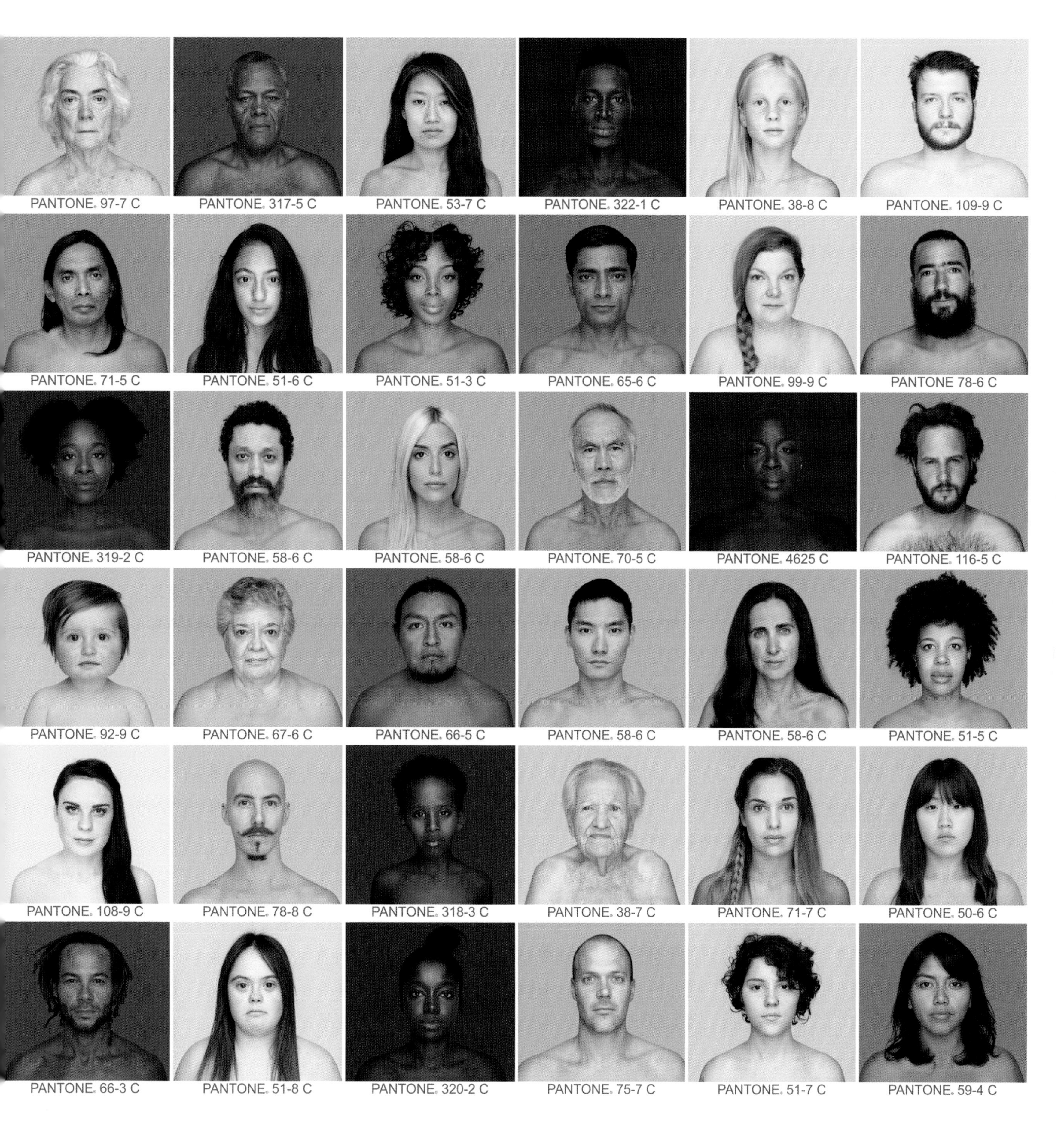
PANTONE® 97-7 C
PANTONE® 317-5 C
PANTONE® 53-7 C
PANTONE® 322-1 C
PANTONE® 38-8 C
PANTONE® 109-9 C
PANTONE® 71-5 C
PANTONE® 51-6 C
PANTONE® 51-3 C
PANTONE® 65-6 C
PANTONE® 99-9 C
PANTONE 78-6 C
PANTONE® 319-2 C
PANTONE® 58-6 C
PANTONE® 58-6 C
PANTONE® 70-5 C
PANTONE® 4625 C
PANTONE® 116-5 C
PANTONE® 92-9 C
PANTONE® 67-6 C
PANTONE® 66-5 C
PANTONE® 58-6 C
PANTONE® 58-6 C
PANTONE® 51-5 C
PANTONE® 108-9 C
PANTONE® 78-8 C
PANTONE® 318-3 C
PANTONE® 38-7 C
PANTONE® 71-7 C
PANTONE® 50-6 C
PANTONE® 66-3 C
PANTONE® 51-8 C
PANTONE® 320-2 C
PANTONE® 75-7 C
PANTONE® 51-7 C
PANTONE® 59-4 C

and into higher latitudes where the sun was weaker, lighter skin was favored. It helped prevent another deficiency—this time, of vitamin D, which is essential for the immune system and normal bone growth; women deprived of vitamin D in their early years are at risk of developing a flattened pelvis, which makes natural childbirth almost impossible. At higher latitudes, a lighter skin color allows the sun to manufacture vitamin D in our skin more efficiently. Our "sepia rainbow" evolved as an evolutionary balance between two health hazards, which mostly depend on where you live: the need to protect our skin against too much ultraviolet radiation and problems created by getting too little.

It was a plausible theory, yet there was no way to be sure how skin color had changed in the past until the revolution in decoding DNA. The first melanin gene was identified in 1999; since then, at least 50 other variants have been found. Different combinations of these variants either turn melanin production on or off, resulting in the subtle rainbow of shades that we see all around us. While darker skin tones are more common near the equator, the frequency of melanin variants in any single population today can be extremely diverse, even at the same latitude; it depends on how long people have lived in a specific region and how much vitamin D the local diet supplies. A recent study of melanin and DNA data from over 1,500 volunteers in Ethiopia, Tanzania, and Botswana underlines the huge variety of melanin combinations from one place to another. "There is so much diversity in Africans that there is no such thing as an African race," says the study's lead author, geneticist Sarah Tishkoff.[138]

Meanwhile, the study of ancient DNA reveals a similar diversity across time. In DNA sequences recovered from Neanderthal bones at two European sites, scientists found a variant that suppresses melanin production, which means that at least some Neanderthals had pale skin and red hair. It makes sense that Neanderthals, who lived in the higher latitudes of Europe and Asia, would have lighter skin than ancient Africans. But the pale skin variant in Neanderthals is *not* found in modern humans today. In fact, the DNA evidence suggests that similar skin colors have evolved multiple times; over the past 70,000 years, Neanderthals, Europeans, and Asians independently evolved light skin using at least three different genes. Similar skin colors have appeared multiple times in human history from completely different combinations of genes, Jablonski says.

Racial categories are meaningless in light of the science that has revealed a riot of genetic diversity underlying the "sepia rainbow." That conclusion is so important in today's divisive world that Jablonski now devotes much of her time to public education about race and the science of human diversity. It is a story that draws on our common evolutionary past, reaching all the way back to when we first left the shelter of the African forests.

LETHAL HEIGHTS: SURVIVING THIN AIR

In 1980, a Buddhist monk climbed up to a cave, sat to pray, and found half of a massive human jawbone with two huge molars lying on the floor. Baishiya Karst Cave is on the steep slopes of Dalijiashan mountain, nearly 11,000 feet up on the fringes of one of the world's most isolated and inhospitable places—the Tibetan Plateau, often called the "Roof of the World." High altitude landscapes—the Tibetan Plateau, the Andean Altiplano, and the Ethiopian Highlands—were the last places on Earth that our ancestors colonized. At 12,000 feet, each breath of air contains only 60 percent of the oxygen molecules in the same breath as at sea level, and hypoxia, or mountain sickness caused by lack of oxygen, is a formidable foe. Many visitors suffer headaches, nausea, sleep problems, and low energy until they acclimatize. Even long-term residents face the hazard of chronic mountain sickness, which can cause overproduction of red blood cells, leading to hypertension, stroke, even death. Yet these hazards were seemingly no deterrent for ancient hunters who visited Baishiya Cave over 160,000 years ago, leaving behind stone tools, cut-marked animal bones, and human remains, including the massive jawbone.

Neglected for decades in a store room at Lanzhou University in China, the jaw was finally studied by archaeologist Dongju Zhang in 2010. No DNA had survived in the specimen, but biologist Frido Welker, supported by The Leakey Foundation, applied a novel technique and successfully extracted collagen, a common protein, from the jawbone's teeth. The ancient protein matched samples from human bone fragments unearthed at Denisova Cave, over a thousand miles away in Siberia's Altai Mountains. (In fact, the jawbone is the *only* fossil of the Denisovans, the extinct side-branch of humanity discovered in 2010, so far known and confirmed outside Denisova Cave.) The DNA results from the jawbone supported the broader genetic picture that Denisovan populations had flourished over a wide area of central and southeast Asia, including the forbidding high-altitude landscapes of the Tibetan plateau. At 11,000 feet above sea level, the Denisovan hunters must have faced thin air, heavy snow, and bitter winds. How did they survive?

As humans first ventured into Earth's high places, each population developed a distinctive set of genetic adaptations to cope with the hazards of hypoxia. Genetic data had previously suggested that this happened for the first time in Tibet around 30,000 years ago, but the discovery of the 160,000-year-old Denisovan jaw suggests that the adaptation may have emerged far earlier.

Some of the most significant evidence of later genetic adaptations in the Himalayas has come from a spectacular source: ancient burial caves exposed by the erosion of sheer 200-foot-high cliffs that tower over the remote villages of Upper Mustang

(Opposite) Sonam is a 12-year-old girl belonging to one of the few remaining families of nomadic sheep and goat herders on the Chang Tang Plateau, which straddles the Tibetan-Indian border at over 13,000 feet. Human adaptation to life at such high altitudes is one of the most striking examples of recent human evolution.

in Nepal. In 2016, a team led by archaeologist Mark Aldenderfer and mountaineer Pete Athans rappelled into the caves to rescue ancient burials, well-preserved by cold and arid conditions yet constantly at risk from the disintegration of the crumbling cliffs. In a precarious operation documented in NOVA's documentary *Secrets of the Sky Tombs*, the team recovered hundreds of human remains, pottery, and ritual objects. Representing three different cultures spanning roughly 1,000 BCE–700 CE, the tomb finds document a variety of shifting burial practices. They range from dismembering and defleshing corpses (similar to Buddhist "sky burial" practices in the region today, in which human remains are fed to wild vultures) to furnishing the dead with little wooden effigies, probably to ward off vampire-like visitations from the afterlife that are traditionally feared in the region.

DNA from the "sky tombs" yielded the genomes of eight individuals, the first ancient human DNA recovered from the Himalayas. Despite the contrasts in rituals from one culture to the next, analysis of the genomes indicates that the identity of people in this remote area has stayed remarkably consistent across 3,000 years, right up to the present day. Among many variations (or mutations) that have helped them adapt to high altitude, two of them, known as *EGLN1* and *EPAS1*, are both connected with reducing the risk of acute mountain sickness by preventing the harmful overproduction of red blood cells.

EPAS1 turned out to have an intriguing origin: the Denisovans, like the owner of the massive jawbone found in Baishiya Cave. The evidence, revealed in a study of the genes of forty living Han Chinese, stunned geneticist Rasmus Nielsen. "There was a complete match," he says. "It's so hard to believe that it could possibly be true. But it is."[139] Nielsen and his team concluded that around 50,000 years ago, Denisovans interbred with the ancestors of today's Tibetans and Han Chinese somewhere in Asia, and passed on their *EPAS1* gene. It had little benefit for the lowland Chinese, since it occurs in only one percent of them today. But for ancient Tibetans, *EPAS1*'s helpful effect on red blood cell production meant that it was favored by natural selection, and is now found in 78 percent of their population.

(Above) Mountaineer Pete Athans rappels into one of the eroding ancient tombs of Upper Mustang in Nepal. Athans was assisting archaeologist Mark Aldenderfer in a precarious rescue operation to salvage and record the tomb contents. (Opposite) The massive jawbone found in Baishiya Karst Cave, one of the few known fossil remains of the enigmatic Denisovans. Genetic evidence suggests that Denisovans contributed a gene that made adaptation to the high altitudes of the Himalayas feasible.

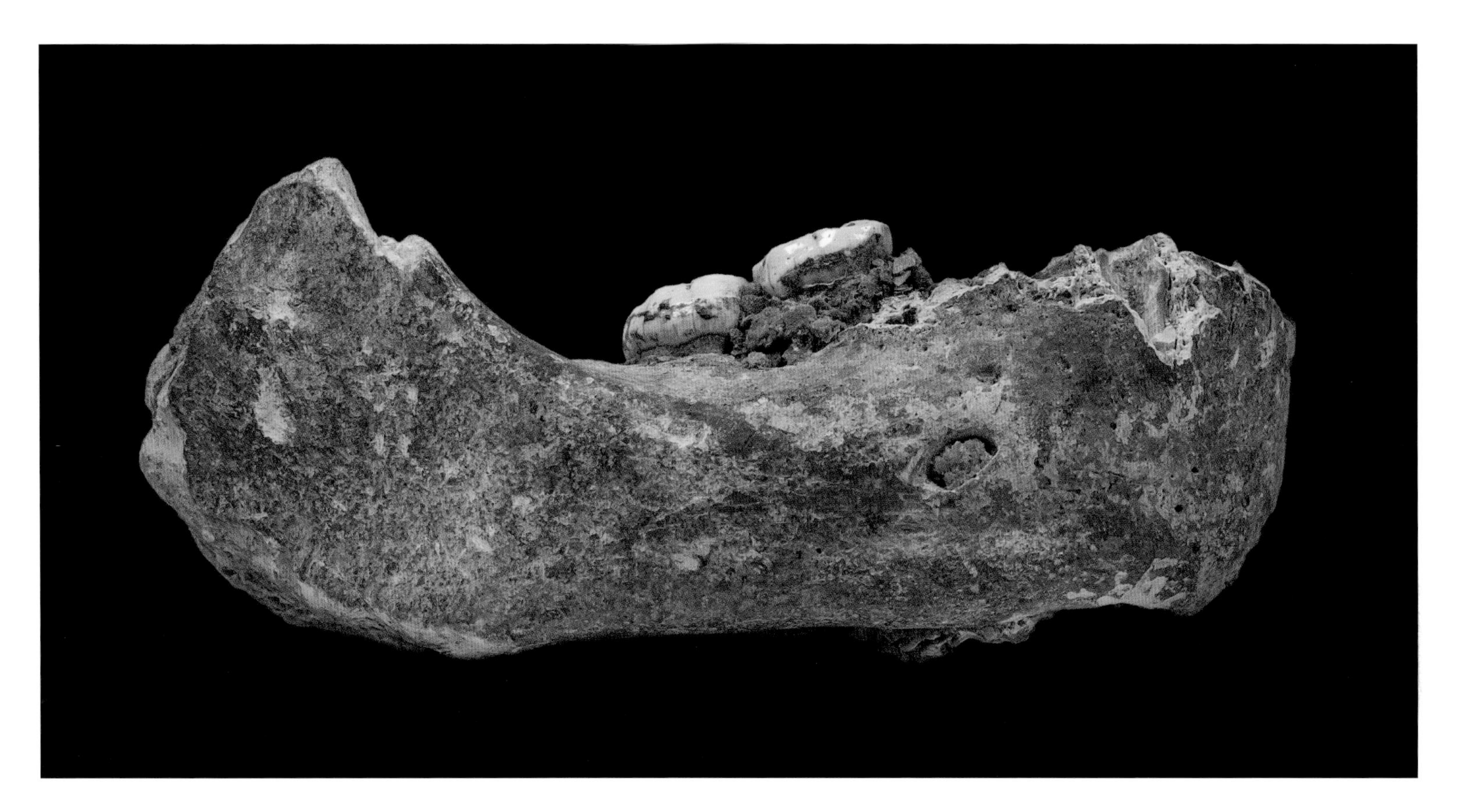

(Unfortunately, no DNA had survived in the Baishiya jawbone to see if it, too, had the *EPAS1* variant.) This finding adds up to one of the strongest and most rapid examples of genetic adaptation yet found in human evolution.

In South America, today's Quechua, the indigenous people of the Andes, have acquired their own set of adaptations to high altitude. Although they share the *EGLN1* gene identified in Tibet, the gene functions differently from its Tibetan counterpart and does not suppress the production of red blood cells; in fact, they have *more* red blood cells ferrying oxygen around the body than people at sea level. As a result, oxygen is delivered more efficiently to their tissues, and babies born to Quechua mothers in the mountains weigh more than European babies born at the coast. Despite that advantage, which promotes infant survival, Andeans run a higher risk of chronic mountain sickness than Tibetans. This suggests that complex trade-offs were involved as the inhabitants of the world's two highest regions adapted differently to the challenges of high altitude.

Supported by The Leakey Foundation, biologist Abigail Bigham and her students are striving to understand the intricate pattern of adaptations among the Quechua, including the question of when and how high altitude genes are "switched on" in the body during childhood. Ultimately, their findings will be of more than academic interest. According to Bigham, "If we can identify specific genetic changes among Andeans that contribute to high-altitude adaptations, they could potentially lead to novel therapies for treating ischemia and other diseases related to the blood supply."[140] If Bigham is right, deciphering high-altitude genes may not only help solve the enigma of how humans colonized the world's most forbidding landscapes; it may point to lifesaving treatments for worldwide diseases that threaten millions today.

DEADLY COMPANIONS

Around 7,000 years ago, a new kind of community moved into the Saxony region of central Germany. They were part of a wave of first farmers, spreading rapidly across Europe from the Near East, the birthplace of agriculture. At the site of Karsdorf, located in the fertile floodplain of the Unstrut River, they raised cattle, sheep, goats, and pigs, and built a succession of sturdy wooden longhouses for their extended families. Alongside one of these houses, they buried a 25–30-year-old man. Recent analysis of the DNA extracted from his bones reveals that he was a victim of hepatitis B, a disease spread by close human contact that today kills nearly 1 million people every year and afflicts more than 250 million worldwide.

The Karsdorf discovery is not only the oldest known case of hepatitis B, but also the oldest viral DNA recovered so far from human remains. The virus's origins, however, almost certainly go back much farther. The team that sequenced the Karsdorf man's viral DNA found that it is most closely related to strains that today appear only in chimpanzees and gorillas. To the team's leader, Johannes Krause, this suggests that the virus could have jumped from ape to human during the early stages of our species' evolution in Africa. "It's more likely this is really an old pathogen in humans for the last hundred thousand years or more," he says.[141] But its presence at Karsdorf 7,000 years ago is significant. Settled farming communities led to rapidly growing human populations; it also meant those human populations lived together in close contact for the first time, providing the virus with many new opportunities to spread.

Until recently, a similar explanation accounted for the origin of tuberculosis, which inflicts an annual death toll of around 1.5 million. Since the TB microbe is a close match for the strain in cattle, a likely theory was that prehistoric farmers caught it from contact with cows in communities like Karsdorf. Alternatively, humans might have infected cattle if the disease emerged earlier, perhaps among our African ancestors. DNA evidence appeared to support this second scenario, since the most diverse forms of the bacterium today are found in Africa. But much remained uncertain, including when the common ancestor of all these strains had first emerged. Rough estimates varied from 3 million to 20,000 years ago.

Then came an unexpected discovery. In 2014, another team led by Johannes Krause and Anne Stone announced the results of an investigation of three 1,000-year-old skeletons from sites in southern Peru. The bones all bore telltale signs of the ravages of TB, notably damaged vertebrae that were abnormally pitted and fused together. The team confirmed this diagnosis by identifying genetic sequences from the tuberculosis microbe in their

(Opposite) A view of hepatitis B, the oldest known virus to infect humans, seen through a scanning electron microscope.

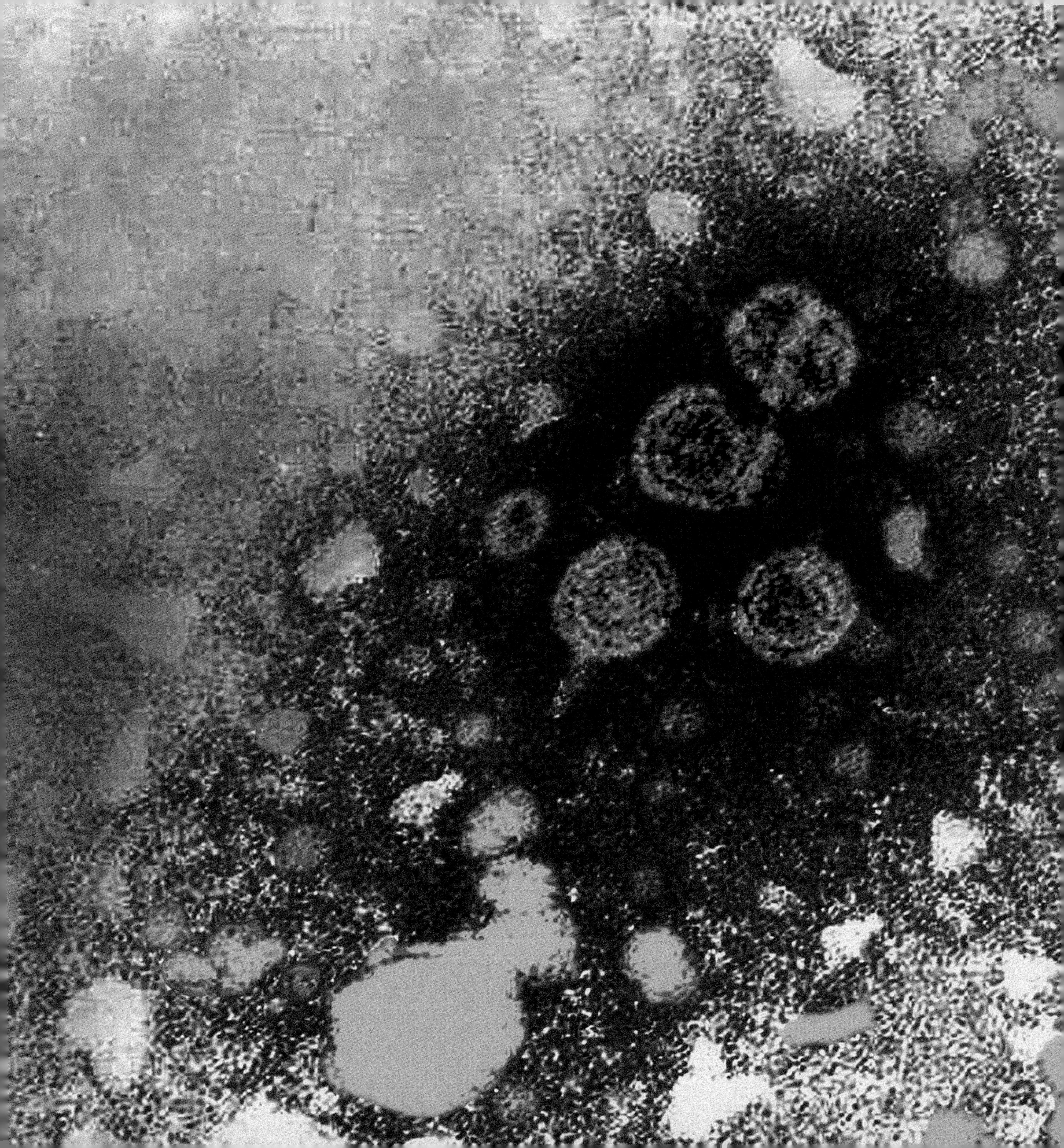

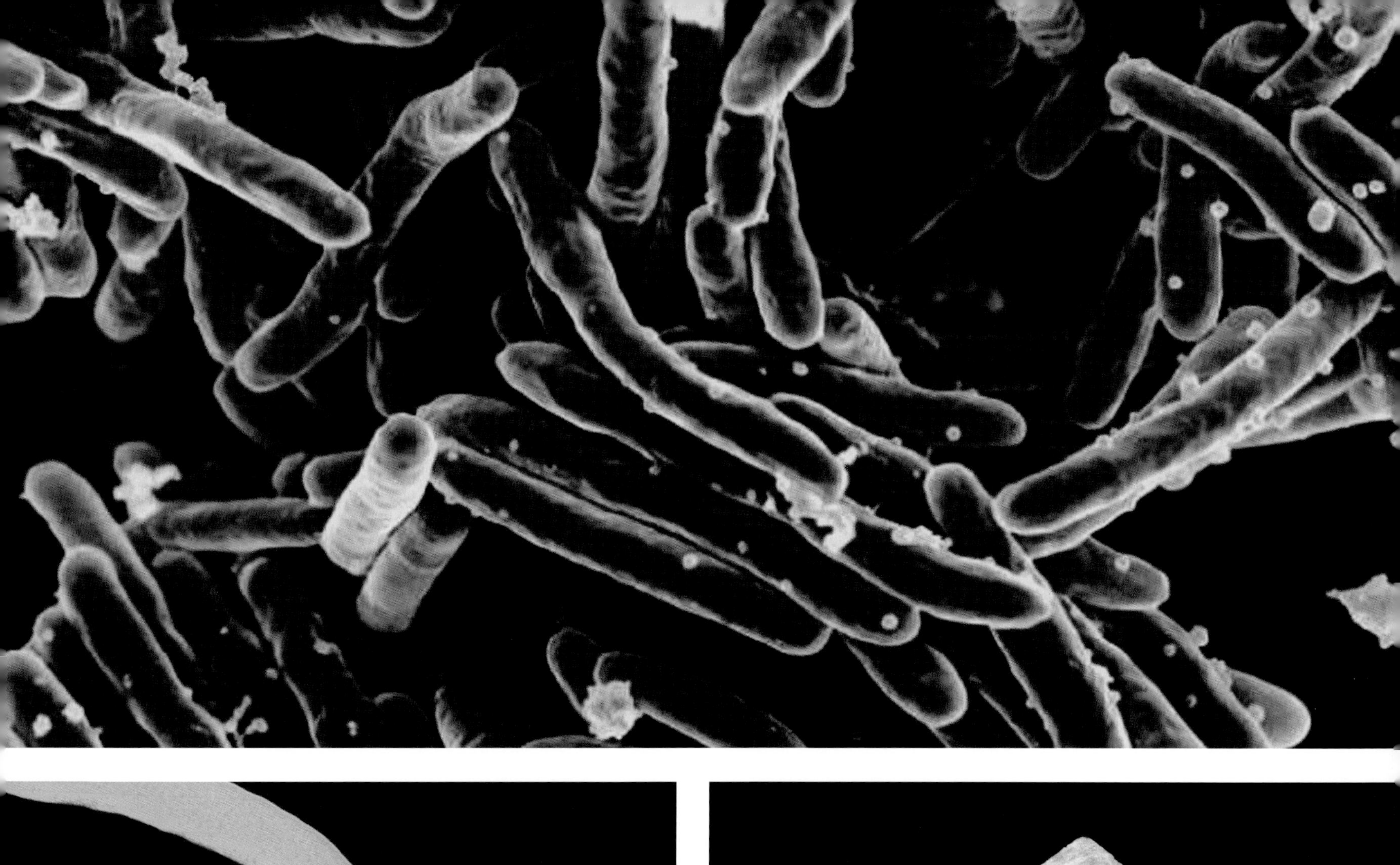

DNA sample. Radiocarbon dating of the bones gave the team a way to fix the age of the bacterium's ancestral origin far more accurately than any previous estimate. To their shock, the result came out at a mere 3,000–6,000 years ago.

This ruled out the conventional scenario of a more ancient emergence of the disease, either among our early ancestors or through the first spread of farming. So where did the TB discovered in the Peruvian skeletons come from, long before Spanish conquistadors arrived to spread European strains of the disease? Further DNA analysis revealed an even more surprising result: the strain was closest to that found not in humans, but in seals and sea lions. Since prehistoric Peruvian coast dwellers regularly hunted and consumed marine mammals, this route of infection is plausible. It suggests a new scenario: after the disease first emerged in Africa, human coast dwellers there spread it to seals, which eventually, through exchange among species, carried it all the way to coastal Peru. While a lot more evidence is needed to reconstruct how tuberculosis spread to other regions, this remarkable discovery shows how today's diseases are the product of an intricate dance between pathogens and their human and animal hosts.

Perhaps the most dramatic example of the co-evolution of disease and humans is sickle cell anemia. In this horrifying affliction, which emerges in childhood, a genetic mutation alters the structure of hemoglobin, the molecule in red blood cells that ferries oxygen around the body. An infant who inherits a single copy of the mutated gene is substantially protected against the most lethal form of malaria. But if a child has two copies of the gene, one from each parent, then red blood cells become deformed and break apart. Symptoms include severe pain, recurrent fevers, strokes, or kidney failure. Some 300,000 babies are born with it each year, mostly in Africa, India, and the Near East. It represents a stark evolutionary trade-off. The mutated gene confers a significant survival advantage to some 20 percent of the population who are silent carriers. Yet that advantage comes at a terrible price, paid by the 1 percent of the population who carry two copies of the gene.

There are currently five major genetic groupings of sickle-cell disease. For decades, biologists have debated whether they stem from one mutation, or multiple mutations that emerged independently, and why some cases of the disease are so much worse than others. One recent analysis traces its origin back to a mutation in a single child around 7,000 years ago in west or in central Africa. During that wet climate period, the child's environment would have been either the equatorial rainforest, or the Sahara region, then a land of woods and savanna, not sand dunes. Both settings would be likely incubators of malaria. If the study is confirmed, it indicates that a single mutation gave rise to all five of today's genetic groupings of the disease. This insight into its evolution may lead to advances in our understanding of the varying severity of sickle cell disease. As in the case of TB and hepatitis B, genetic evidence is expanding knowledge of the patterns of ancient disease, holding out the hope of improved treatments in the future.

(Opposite, top) *Mycobacterium tuberculosis*, which causes TB. (Bottom left) One of the deformed red blood cells that create the devastating symptoms of sickle cell anemia. (Bottom right) The 7,000-year-old skull of a young male from a Neolithic settlement at Karsdorf in central Germany. DNA from the skull revealed the oldest known case of hepatitis B.

THE AFRICAN MILK TRAIL

For months on end, biologists Alessia Ranciaro and Sarah Tishkoff traveled hundreds of miles to remote African villages to carry out perhaps the most ambitious genetics study ever performed in Africa. Their goal was to tackle one of human evolution's most intriguing puzzles: why some people, but not others, are able to digest milk. More than a mere curiosity of dietary history, the solution turns out to be one of the most dramatic examples of how cultural and genetic changes are sometimes intertwined and evolve together.

After obtaining ethical consent for their study, Ranciaro and Tishkoff would contact local authorities and tribal chiefs on arriving in a village and convene a public meeting to recruit volunteers for their study. Participants had to fast overnight, then give a series of blood samples next morning after swallowing a strange orange-flavored drink. "This was a very challenging test to do in the field in remote regions," says Ranciaro. "We were careful to make sure that people understood why we were doing this study and that they would need to commit to the hour or more of time needed to do the test."[142] By the time their research was published in 2014, they had sampled over 800 people from more than 60 different populations in Kenya, Tanzania, Sudan, and South Africa.

The biology behind the milk puzzle is well understood. All humans digest their mother's milk as infants, but after weaning, many lose the ability to produce the enzyme lactase, which is essential for breaking down lactose, the sugar in milk. Until cattle were domesticated around 9,000 years ago, we no longer needed to digest milk after weaning. Today, for nearly two-thirds of the world's population, drinking cow's milk will result in an urgent trip to the bathroom. The puzzle is why some of us developed the ability to continue to produce the enzyme lactase as adults—"lactase persistence"—and so be able to down pints of milk without discomfort.

From the 1940s, the health benefits of drinking milk were increasingly promoted in the U.S. and northern Europe, where lactase persistence is common. When researchers found that many African Americans and Asian Americans lacked the enzyme, the deficiency was cast in racial terms; a 1968 *New York Times* headline read: "Most Nonwhite Persons Found to Develop an Intolerance to Milk." However, anthropologists soon reported examples of lactose tolerant pastoralists who rely heavily on milk and dairy products in many parts of West, North, and East Africa as well as the Middle East. For the Maasai and other cattle herders in Kenya and Tanzania, consuming milk is central to their lives and celebrated as a daily blessing.

In 2002, a team of Finnish biologists identified a mutation in a gene associated with the production of lactase in adults. Not

(Opposite) Kamaika Kingi is a 33-year-old Maasai herder from southern Kenya. The traditional Maasai diet consists mainly of the milk, blood, and meat of the cattle that are their source of wealth.

surprisingly, its distribution closely matches the pattern of milk drinking: it is commonest in the U.S. and northern Europe, and gradually tapers off across the rest of Eurasia. When scientists analyzed the data to look for the likely age of the mutation, they estimated it had first appeared sometime between 2,000–21,000 years ago. This was broadly consistent with the theory that it was linked to the domestication of cattle and the spread of farming from the Near East across Europe after 9,000 BCE. According to the theory, the ability to digest milk opened up a rich new source of nutrition that gave prehistoric dairy farmers a significant survival advantage, perhaps as a fallback in times of famine or drought.

But this explanation had one big problem: the Eurasian mutation was absent in nearly everyone in Africa and the Middle East who drank milk. So, in the early 2000s, Sarah Tishkoff launched her first field expedition to discover if different lactase-regulating mutations existed among rural cattle herders in Africa. By the time her work culminated in the 2014 study, her team had identified no less than four new mutations, each one enabling specific populations in particular regions to digest milk. Evidently, each mutation had arisen independently of all the others—a striking case of convergent evolution.

More remarkable still, when the team investigated the distinctive stretches of DNA, or haplotypes, associated with each mutation, they found that natural selection had favored the African ones in relatively recent times—between about 3,000 to 7,000 years ago. These dates correspond roughly to the domestication of cattle in Africa that began around 6,000 years ago. Moreover, the selection signal is phenomenally strong; of her first African results, Tishkoff said, "It is basically the strongest signal of selection ever observed in any genome, in any study, in any population in the world."[143] The implication is that milk drinking offered such a powerful survival edge that nature kept rapidly "inventing" new versions of the lactase-enabling mutations wherever cattle raising was adopted.

The story of milk underlines the fact that, contrary to widespread popular belief, our bodies did not stop evolving when we became "modern humans." In reality, cultural innovation can reshape our genetic identity in the blink of an eyelid, from an evolutionary perspective; and in the view of most scientists, the pace of human evolution is accelerating with the explosive growth of migrations and megacities. Meanwhile, Sarah Tishkoff's far-ranging DNA studies of African populations are revealing the extraordinary diversity and richness of Africa's genetic heritage, which scientists have only just begun to tap.

(Opposite) In a study carried out by Alessia Ranciaro and Sarah Tishkoff, the pastoralist Hamer people of southwestern Ethiopia were tested for their ability to digest milk. A gene facilitating lactase digestion was strongly favored by natural selection as cattle herding spread across Africa a few thousand years ago.

THE STRANGE RITUALS OF CAPUCHINS

Primatologist Susan Perry was several years into her study of capuchin monkeys when she began observing some highly peculiar behavior. Capuchins are small, agile, feisty, and inquisitive primates, common to forests in many parts of Central America and Brazil. At the time of her first trip to Costa Rica in 1990, everyone assumed that like other New World primates, capuchins led relatively simple social lives. Instead, Perry found a hotbed of seething political tensions and status rivalries in her Costa Rican groups, and what was to have been a short visit has become, more than a quarter of a century later, one of the most comprehensive field studies of any primate. Supported by The Leakey Foundation, the resulting data add up to a rich picture of the intricate web that binds capuchin society.

Perry was drawn to study capuchins by their reputation as "the brainiest monkey,"[144] known for their clever manipulation of tools and creative foraging; a lot of the time, they are in constant motion, searching for fruits with their friends or squabbling with their foes. But every now and then, a pair of them will sit down face-to-face with what Perry calls "a Zen-like look,"[145] and lock one another into strange rituals seemingly designed to test their patience and tolerance for pain. Over more than a decade at a remote field site in Costa Rica, Perry's team meticulously recorded a handful of such rituals: "hand-sniffing," in which both monkeys insert their fingers into each other's nostrils simultaneously for prolonged periods; "sucking body parts," mutual sucking on partners' ears, fingers, or tails for minutes at a time; and most bizarre of all, "eyeball-poking." This entails one monkey slipping its finger between the eyelid and the bottom of the eyeball up to the first knuckle; in this painful embrace, they remain frozen in a trance-like state, "and often the one being poked in the eye inserts fingers in the partner's nostrils or mouth during the eyeball-poking," notes Perry.[146] Based on 20,000 hours of field data, Perry found that each ritual was invented by individual monkeys, then caught on like a fashion or fad within the group, often with slight variations from one pair to the next, before eventually dying out. What could explain such odd behavior?

In the capuchin world, females bond together strongly but power and influence are mainly wielded by the alpha male. When it comes to mating, it looks like a level playing field—all males get an equal chance to mate with all females. But Perry's long-term DNA data revealed a big surprise: when it comes to what really matters—producing offspring—the alpha male is responsible for over 90 percent of the breeding in the group. Since some alphas can hold on to power for as long as 18 years, the result is a growing number of sons and daughters that can help support his rule.

(Opposite) A pair of white-faced capuchin monkeys in northwest Costa Rica engage in "hand-sniffing," one of a half-dozen bizarre behaviors apparently intended to reinforce trust and test alliances in the capuchins' highly competitive society.

Despite the alpha male's dominant role, his position is always insecure. "Getting to year six or seven is the really difficult part," says Perry. "After that, their sons are old enough and they can coast for a while, since sons are intensely loyal; we've never seen them turn against a father."[147] But brothers are a different story. "We see intense sibling rivalry reminiscent of the British monarchy's issues. We've had brothers kill brothers for the alpha male position. And to take over and conquer a neighboring group, building alliances is essential."[148]

Susan Perry describes how one such takeover figured in the life of an "unusually large and greedy" male capuchin that the team called Moth. When he was ten, he began visiting other groups with his favorite playmates, Newman and Tranquilo. A year later, Moth and Tranquilo successfully invaded one of these neighboring groups; to consolidate his power, Moth turned against Tranquilo and evicted him. Now he was the alpha, but he never ingratiated himself with the females he ruled, and after killing one of their infants, he was himself ejected and returned home. Back in his original group, he allied himself with the elderly alpha, Pablo, who wielded political expertise, while Moth, according to Perry, was mainly interested in eating but "contributed physical formidability." Their co-dominance worked for a few years until Moth deposed Pablo in a relatively gentle fight. Pablo, who had sired 25 children, stepped down and stopped breeding. For the next six years, Moth ruled largely unopposed, until he went missing during a severe drought in 2014–15.

The turbulence of capuchin politics presents the most plausible explanation for the strange rituals that Perry's team observed. Since strong alliances are vital for survival, one way to test them is to engage in these ritualized encounters. Each one imposes an element of discomfort and risk, providing a far clearer signal of trust than the simple act of grooming that helps to cement relationships among other primates. The rituals provide a way of enhancing the capuchins' sensitivity to the past and present behavior of their allies. As Frans de Waal concludes: "What these monkeys seem to be telling each other—similar to humans when they drop backward into the arms of others—is that based on what they know about each other, they have faith that all will end well."[149]

(Opposite) Primatologist Susan Perry (in the background) studies the complex traditions and intricate dynastic struggles of capuchins. Her team uses DNA analysis to figure out relationships and genealogies, revealing, Perry says, "some pretty intense sibling rivalry reminiscent of the British monarchy's issues."

KILLING THE INFANTS

In the summer of 1971, Sarah Hrdy, a 25-year-old Harvard graduate student, traveled to Mount Abu, a steep, lushly forested peak rising from the arid plains of northwestern India. She was intrigued by reports that the local male langur monkeys occasionally killed infants, a behavior blamed on the stress of overcrowding as human communities expanded into their habitats. While an undergraduate, Hrdy had attended a class taught by biologist Paul Ehrlich, author of the best-selling *The Population Bomb*, that warned of the risks of overpopulation. She hoped that the langurs' abnormal behavior might serve as a case study for understanding the effects of overcrowding on humans. Instead, over five summers and hundreds of hours of langur watching, Hrdy developed a provocative theory that infanticide was far from abnormal; in fact it was an adaptive strategy, enabling male monkeys to father as many offspring as possible. Coinciding with the emerging, controversial new field of sociobiology, her theory ignited fierce controversy. At that time, it was widely believed that animals behaved to enhance the survival of the group, so how could such individually selfish behavior evolve? Could the seemingly ruthless killing of infants have contributed to humanity's evolutionary success?

Langur—or "leaf"—monkeys are remarkably resilient and adaptable. They thrive in a wider range of habitats than any other primate, from India's parched plains to 14,000-foot slopes of the Himalayas, where they are occasionally mistaken for the fabled Yeti. A centuries-old Hindu legend casts these agile, black-faced langurs in a heroic role battling evil forces, which ensures that they are protected and fed even in some of India's major cities. On Mount Abu, Hrdy studied five troops on the edge of town, consisting either of a single adult male together with their closely related females and infants or of small, roving bands of unattached males. Despite crowded conditions, the langurs' behavior confounded her expectations. "Here were these infants playing around," Hrdy remembers, "bouncing on these males like trampolines, pulling on their tails, and so forth. These guys were aloof but totally tolerant. They might show some annoyance occasionally, but there was nothing approaching pathological hostility toward offspring. The trouble seemed to be when males came into the troop from outside it."[150]

In the case of one small langur group, Hrdy recorded separate takeovers by at least four different outsider males. Each time there was a switch of leadership, one or several infants would disappear. On three occasions, she witnessed the incoming male launch fatal attacks. By 1974, her data indicated that infant mortality had soared to over 80 percent, making it likely that this small Mount Abu group would soon die out.

One of Hrdy's most crucial observations was that an attacking male always targeted unweaned infants in the group he was attempting to take over. As long as a langur mother nurses her young, she cannot conceive; in some troops, she remains unavailable for as long as two years. But if her infant dies, she becomes receptive again in only about eight days, and will often give birth some six to eight months later. Here was a compelling reason for the pattern of male violence. By eliminating a nursing infant and mating with the bereaved mother, an incoming male could speed up her reproductive cycle and maximize his chances of siring her next infant. Moreover, Hrdy found that dominant males typically hung on to power for only about 27 months before being driven out by an outside rival. If he were to spare the existing young and wait for the female to become available during his relatively brief tenure, then his own infants would be unweaned and vulnerable by the time a new usurper came along to challenge his rule. In this highly competitive situation, Hrdy says, a new male could "compress the mother's

The society of langur monkeys in northwest India is generally peaceful, but when they stage hostile takeovers of rival groups, it often leads to a shakedown of ruling alpha males and infanticide.

fertility into the period he was likely to have reproductive access to her before another male replaced him."[151]

Hrdy's theory received a strong boost in 1999, when primatologist Carola Borries sampled the DNA of langur groups in southern Nepal. In every case of a baby killing or attack, the DNA showed that the males were unrelated to their victims. When the victim's mother later gave birth to another infant, the killer was invariably the father. Meanwhile, evidence has steadily grown that a similar phenomenon is common among creatures as diverse as mice, bees, bears, hippos, squirrels, wasps, and dung beetles. The lions of Africa's Serengeti park, well-documented over 30 years, offer a particularly stark example. During times of stable leadership by a dominant male, over 90 percent of lion cubs survived their second birthday. But when an outside male takes over, only 14 percent lived beyond nine months. The killings are methodical. "Males are not in a hyperkinetic frenzy," notes biologist Craig Packer, "and the cubs are not accidentally caught in the middle. The males are focused; the cubs are the targets. In lions, infanticide is one of the fundamental facts of social life."[152] Among the langurs, Hrdy observes, male attacks are "as organized and focused as a shark's."[153]

Did this type of infanticide shape the lives of our early hominin ancestors? One clue stems from the fact that in many primate species in which competition for females is fierce and infants are at risk, males are typically much bigger than females. In this situation, larger, stronger males are more likely to reproduce, either because females select them or because they can more readily defeat their rivals. Although the data are controversial, measurements of fossil bones suggest that there were strong differences in height and body weight between males and females among some early human ancestors, including the australopithecines. Among Lucy's species, for example, males appear to have been twice the size of females—a hint that aggressive competition for females and lethal attacks on infants may have figured significantly in our evolutionary past.

(Opposite) Sarah Hrdy observes langur monkeys at Mount Abu in northwest India during the 1970s. Hrdy's pioneering conclusions about infanticide among langurs led to her highly influential studies of motherhood in human and primate society.

At Mount Abu, Hrdy discovered that langur mothers are far from helpless in the face of threats to their infants from outsiders. Even when a female is not actually ovulating, she will often seek to mate with outsides males, soliciting sex by "faking" her estrus behavior: presenting her rump to the male and frenetically "shuddering" with her head. If the male is aroused and convinced by the performance, he takes up the offer. While some researchers found these reports of female promiscuity shocking, Hrdy interpreted it as a defensive strategy, designed to protect the infant she would eventually deliver by sowing confusion about its paternity. Since males never attack the infants of females they have mated with, "what from a male view looks like promiscuous behavior," she comments, " is from a female's view assiduously maternal."[154]

Since Darwin, scientists had taken it for granted that males would be eager to mate with any female they could, while females tended to be far more discriminating and "coy." So Hrdy's ideas about the selective pressures that led females to mate with multiple partners were regarded as heresy. But as examples of other species with similar evidence of selection pressures on both sexes poured in, acceptance of Hrdy's provocative work among the langurs work grew. It laid the foundation for decades of her highly influential investigations of female sexuality, mothering, and infant care. Her langur research had paved the way by showing how females could manipulate their sexuality to manage the male threat and insure their reproductive future.

MONKEY MONOGAMY

In 2018, one of Britain's most prolific mothers, Sue Radford, gave birth to her 21st child at the age of 43 amid a blaze of tabloid publicity. She captured attention because the physical demands of pregnancy, birth, and nursing, together with the onset of menopause, severely limit the number of children a woman can have. No such limitations affect men. For instance, in 1704, French diplomat Dominique Busnot visited the court of the infamous Emperor of Morocco, Moulay Ismael the Bloodthirsty. If Busnot's report is true, he may have sired over 1,000 children from four wives and 500 concubines over 32 years—a claim that anthropologists recently tested with a computer model and concluded was, in principle, biologically possible. Even if we discount history's tales of relentless inseminators, the reproductive potential of men and women is vastly different. Since men can pass on their genes to multiple partners at any time, why is monogamy so widespread in human society? Even in countries where polygamy is lawful, a majority of the population still tends to pair off into long-term relationships. The pervasiveness of enduring romantic attachments suggests that monogamy—or "pair bonding," as biologists prefer to call it—has deep roots in the human story, and is somehow fundamental to our species' success. But how did it evolve?

The puzzle grows more complicated if we consider that pair bonding is common in birds but rare in mammals, appearing in fewer than 10 percent of all mammal species and around just a quarter of primate species. Moreover, those figures apply only to social monogamy, in which couples live together and raise their young but also potentially have multiple partners. Far rarer still is genetic monogamy, in which couples have exclusive mating relationships. That has been reported in only a handful of mammal and primate species—among them, one of South America's most unusual and intriguing monkeys.

Azara's owl monkeys live in South America's Gran Chaco, a vast expanse of patchy forests and grasslands, nicknamed "Green Hell" for its mosquitoes and extreme temperatures. In 1996, newly minted Ph.D. Eduardo Fernandez-Duque relocated his family to a remote cattle ranch in the Argentinian Gran Chaco to begin more than two decades of intensive fieldwork. Drawn there by the pair bonding puzzle, and supported by The Leakey Foundation, he soon realized how challenging his project would be. Tree-dwelling owl monkeys are small and inconspicuous, and both males and females—in fact, all individuals—look almost identical. For three years, his research team struggled to identify specific animals, an essential task if they were to understand the basis of pair bonding. Only with the help of radio collars and receivers were they finally able to make progress in recognizing and tracking monkey couples, following them in 4–8-hour shifts both day and night. Fernandez-Duque says he was determined "to study enough animals to acquire a baseline of data that can inform our reconstructions of the evolutionary past. We have to get solid data on living primates if they are to become a strong model for interpreting the fossil record."[155]

After half a decade of year-round observation, a remarkable picture of owl monkey social life began to emerge. Each group includes a pair-bonded couple and their young, who spend most of their time around a core feeding area, surrounded by a broader

(Opposite) A mother, father, and two infant owl monkeys huddle together on a cold morning in the Argentinian Gran Chaco forest. Eduardo Fernandez-Duque and his colleagues investigate the riddle of why owl monkeys are one of the few monogamous primates besides humans.

(Opposite) Owl monkeys look almost identical and are hard to spot in the forest. Cebollita, the 20-month-year-old daughter from the owl monkey family shown on p. 236, wears a radio tracking collar around her neck. This technology proved essential to figuring out relationships within owl monkey communities.

territory that they aggressively defend against outsiders. "The pairs we observe are really very coordinated," Fernandez-Duque says. "They move together, feed together, and raise their young together. The males are extremely committed to paternal care, often playing and carrying the infants, while female interaction is mainly limited to nursing. We've never seen divorce and never seen infanticide."[156] DNA testing of couples and their infants has so far shown an extraordinary record of faithfulness: owl monkeys are the first primates proven to be genetically monogamous.

The project's extensive field data provide a rich source for testing the many theories about why pair-bonding originated. One likely explanation lies in the spotty distribution of fruit trees in the Gran Chaco's relatively open forest, which in any given area can support only a single female rather than large groups. This wide dispersion of females makes it impossible for males to control more than one female at a time, and helps explain why both males and females defend their territories so aggressively. Another theory highlights the benefits provided by paternal care. The presence of devoted fathers boosts the chances of infant survival and means that females can recover more quickly from each pregnancy. While these advantages are obvious once monogamy is in place, it remains such a rarity among mammals that it is still unclear why it originates in the first place.

Fernandez-Duque found support for both theories in the owl monkey data, yet remained convinced that other factors are also at play in their unique behavior. After several years of intensive monitoring, the team began reconstructing a vital missing piece of their life histories—one that paints a less rosy picture of their social life. Around the age of three, both males and females leave the security of their home groups and become "floaters," lone individuals who wait cautiously around the edges of other territories, looking for a chance to acquire a mating partner by breaking up an established couple. Since owl monkeys fiercely defend their territories and mates, the process involves dramatic fights between rivals, sometimes with fatal outcomes. During the first decade of observations, the team recorded fifty cases of "replacements," in which a male or female floater successfully expelled and replaced an existing mate of the same sex. "These are high-stakes competitions for reproductive positions," Fernandez-Duque notes. "Only once have we seen an animal kicked out of a group move successfully into another group. Often, after they're expelled, they're wounded and don't make it...we don't see them anymore."[157]

Paradoxically, then, the devotion of owl monkey couples goes hand-in-hand with intense sexual competition. They turn out to be serial monogamists, fiercely loyal to their partners until a floater intrudes and breaks them up. For those forced to take on new partners, there is a price to be paid. The data shows that, on average, they have 25 percent fewer babies than couples who manage to repel outside attacks and hang on to an exclusive, enduring relationship until one partner dies. The reproductive edge of these long-term couples may be another factor that explains the persistent foothold of pair bonding among owl monkeys.

We may never know exactly which explanation accounts for the rise of monogamy among our ancestors, but Fernandez-Duque's years of painstaking work have opened up unexpected new perspectives on this deep-rooted aspect of our evolutionary past. As he observes, "Call it love, call it friendship, call it marriage, there is something in our biology that leads to this enduring, emotional bond between two individuals that is widespread among human societies."[158]

MAPPING THE GENOME: OUR UNFINISHED JOURNEY

Our evolutionary journey, which began with tiny, one-ounce primates in the forests of Asia 55 million years ago, was nothing like the straight-line march of progress visualized by scholars and scientists half a century ago. Instead, like a braided stream, our ancestry has wound an intricate course, dividing many times since the hominin line split off from that of the great apes, then rejoining and mixing again, and sometimes petering out in dead-end rivulets.

Even when anatomically modern humans emerged in Africa, their survival was far from pre-ordained. Around 100,000 years ago, climate fluctuations and other factors may have reduced their population to as little as a few thousand. But these same abrupt climate swings, and the changing habitats that resulted, are what forged our ancestors' unique adaptability.

Mastering the production of tools, protective clothing, and watercraft allowed us to colonize every corner of the globe. We thrived in a vast range of environments, from the Himalayas to the Arctic to the Kalahari. We survived the rigors of the Ice Age, when we co-existed with at least four other species—Neanderthals, Denisovans, the Hobbit of Flores, and South Africa's *Homo naledi*—all of whom ultimately died out, leaving us as the last remnant of the hominin family.

Along the way, we have shared our evolutionary journey with those of other primate species, currently numbering over 500 and many critically endangered. The study of our primate relatives illuminates many alternative pathways that early human behavior and adaptation might have taken.

Over the last 10,000 years, our most powerful means of adaptation—human culture—enabled us to take ever-greater control of our circumstances and the natural hazards we faced. Our forebears helped drive giant Ice Age mammals into extinction, create new food resources by farming, reshape landscapes, and, after less than three centuries of industrial activity, alter the climate itself.

Now, at least in much of the developed world, enormous advances in healthcare have lengthened lifespans. With sanitation, adequate housing, and medical care, women live long beyond their reproductive age and the majority of children survive into adulthood. Ailments that were once potentially fatal such as appendicitis, dysentery, or anemia are now easily cured. Genetic testing offers us the chance to detect lethal conditions in the womb. Gene editing tools like CRISPR open up novel possibilities for directly altering our DNA. These advances promise a new era of personal precision medicine but also pose ethical risks. Insulated against many of the once-formidable forces of natural selection, it is tempting to think that we have seized the reins of our evolutionary destiny.

(Opposite) As this recreation from NOVA's *Great Human Odyssey* evokes, modern humans proved to be highly resilient and adaptable as they left their tropical African home and populated every corner of the globe.

One widely shared assumption in popular culture is that human evolution has stopped, and that our biological identity is frozen at the hunter-gatherer stage of our existence. That is certainly the notion of Paleo-diet enthusiasts, who think that our metabolism and digestive system has still not yet evolved to cope with grains, cereals, and dairy products. A leading proponent writes that, since "the genome makeup of Paleolithic people is virtually identical to our own...we are returning to the diet we were genetically programmed to eat."[159]

Another widely held belief is that cultural forces have outpaced natural selection. Even some well-known biologists such as Steve Jones have argued that, in the developed world at least, "natural selection has to some extent been repealed...things have simply stopped getting better, or worse, for our species."[160] In an interview, the late Stephen Jay Gould said that natural selection has been so outstripped by the speed of cultural change that it "has almost become irrelevant in human evolution. There's been no biological change in humans in 40,000 or 50,000 years. Everything we call culture and civilization we've built with the same body and brain."[161]

Gould's remark overlooks many physical changes that have reshaped our bodies and brains during this timespan. On average, we are shorter, lighter, and less ruggedly built than modern humans were during the Ice Age. The advent of farming has made us dependent on a narrower, less healthy diet of staple crops, and brought new diseases spread by domesticated animals. As a result, by some estimates, the average height of European men at one point had fallen by eight inches. Only in the last century or so has the height trend reversed, thanks mainly to modern improvements in diet and healthcare. Our brains, however, are still smaller than those of early modern humans. Under the influence of a softer diet, our jaws and teeth have also shrunk, as anyone who has had their third molars (or wisdom teeth) extracted because their jaw lacks space, is painfully aware.

Natural selection is not the only way that genes change from one generation to the next: they are also shifted by a variety of chance events, known to biologists as genetic drift, and also by gene flow, when populations mix as they move into new regions or environments. In addition, random mutations (or "copying mistakes," generated every time a cell divides) mean that our genome is constantly in flux. On average, each of us have around 150 new mutations that our parents did not have. Moreover, global populations have exploded by over a thousandfold since the dawn of agriculture, when there were perhaps only 5 million people. A growing population multiplies the number of new mutations present in any group. That means a vast increase in the number and variety of potential targets for natural selection—some beneficial, some potentially harmful—and a growing likelihood that rare variations will survive and pass on to the next generation. This changeability underwrites the incredible diversity of body build, skin color, hair, and other attributes so evident in human populations today.

Not until the first large-scale genetic studies could we comprehend the full measure of how much our biological identity has shifted over recent millennia. Far from being "frozen" at the dawn of agriculture, the rate of DNA change, and hence the pace

(Opposite) The nearly intact skull of a 15- or 16-year-old girl was discovered in 2007 by divers at the bottom of Hoyo Negro, a flooded limestone cave in the Yucatan, Mexico. DNA from the 13,000-year-old skull, one of the earliest in the Americas, helped confirm that today's Native Americans are all related to a single founding ancestral population.

DUI
TLS

Safe Imager™

(Opposite) Sequencing DNA has transformed every aspect of studying the human past and how we relate to our primate cousins. Here, Leakey grantee Mareike Janiak prepares DNA samples in a University of Calgary lab to identify digestive enzymes in primates.

of human evolution, is drastically speeding up. A 2007 study reported on distinctive "single-letter" genetic markers (known as single nucleotide polymorphisms, or SNPs), drawn from 270 people in four different populations. Based on this sample, the researchers estimated that the pace of human evolution is now 100 times faster than it has ever been since the earliest hominins split off from the great apes. They attributed this accelerated pace to the rising tide of cultural innovations, sweeping lifestyle changes, and explosive population growth since we took up farming.

By 2010, another transformation in our understanding was under way. DNA sequencing and massive advances in computing power meant that researchers could begin to detect evolutionary change not merely in single marker genes, but across entire genomes. The scale of many of these new studies was astonishing. In 2013, geneticist Joshua Akey and his team compared the genomes of 6,500 people from Europe and Africa, scanning 15,000 genes and detecting more than 1 million variations in the sampled DNA. No less than three-quarters of them turned out to have arisen over the last 5,000 years. In other words, the majority of changes in our genome have occurred relatively recently, and we carry a far greater amount of mutations than our ancestors did. "The genetic potential of our population is vastly different than what it was 10,000 years ago," says Akey. "In a way, we're now more evolvable than at any time in our history."[162]

Most mutations turn out to have little or slight effects in the body. Others can be harmful if they alter proteins and make them defective, potentially creating new ailments or contributing to disorders that are now prevalent in the modern world, notably diabetes, obesity, and insulin resistance. But if new mutations arise with strongly beneficial effects, they can spread rapidly through the population in only a few hundred generations. Such was the case with lactase persistence, a mutation in a specific gene that prolongs the ability to digest milk beyond infancy (see p. 225). First emerging in southwest Asia around the time of the earliest domesticated cattle around 9,000 years ago, the ability to tolerate milk gave dairy farmers an enormous advantage in overall health. It enabled mothers to bear more children and therefore ensured that the genetic variant would spread rapidly. The effect was so great that different lactase alleles appeared independently at least five times across Africa, Europe, and Asia, and represent some of the strongest signals of natural selection ever detected in humans. Sickle cell anemia is another classic case of a fast-spreading mutation. A single genetic "letter" change in an African child 7,000 years ago offered protection from malaria. Yet it imposed a deadly cost of defective red blood cells on a minority of the child population (see p. 223).

Apart from a handful of impressive cases, however, it is increasingly clear that such "one-letter" gene changes, spreading like wildfire through a population, are relatively rare. Many other traits spread more slowly and involve hundreds, if not thousands, of genes acting together in concert. In 2014, an aptly named international team called GIANT drew on massive computing power to search for genes associated with human height differences. The data consisted of an enormous sample of over a quarter of a million people, mostly Europeans. The results indicated that height is affected by nearly 700 genes scattered across more than 400 different spots in the genome. Each gene by itself has little effect, usually affecting stature by only a millimeter or two, and no single person has all of these gene variants, or they would indeed become giants. Instead, selection pressures orchestrate different combinations of genes to act together and produce all the variety of heights we see in populations today. A recent new analysis of the GIANT data calculates that the total number of gene variants affecting height could eventually prove to be more than 100,000.

If regulating height, a simple physical trait, involves thousands of genes, a much steeper challenge confronts researchers striving to disentangle the genetics of complex conditions such as autism, depression, or schizophrenia. The intricacy of gene networks is one reason why progress in developing gene therapies for chronic diseases such as cancer has been so slow. Moreover, since individual genes often have multiple functions, tinkering with any one of them may have unpredictable results.

For the same reason, engineering genes to manipulate attributes like intelligence or creative ability remains a difficult and dubious prospect, practically and ethically. But recently, a new genetic technique has seemingly brought the era of "designer babies" closer. Non-invasive prenatal testing (or NIPT) involves analyzing tiny amounts of fetal DNA that circulate in the blood of an expectant mother. While providing a safe, early way to screen for Down syndrome and rare birth defects, such tests can also detect sex and, in principle, could be used to signal traits like height, hair color, or the risk of future disease. Ethical constraints may eventually prove weak against pressures to correct undesired traits or add protective advantages. Experiments in creating "designer babies" may come all too soon, as the recent use of the editing tool, CRISPR, to insert an HIV-protective gene in a pair of twins in China warned us.

The promise and perils of our genetic future are beyond the scope of this book, as are even more speculative scenarios involving our deepening reliance on computing devices and the future role of Artificial Intelligence in humanity's destiny. Echoing many prophets of A.I. doom, biologist Peter Ward imagines that "advanced artificial intelligence could encapsulate the various components of human cognition and reassemble those components into something that is no longer human—and that would render us obsolete."[163]

One thing is certain: our evolution has not stopped. While our technology and culture give us increasing power over the natural world, the forces of natural selection are not done with us.

"Humans haven't really changed the rules of natural selection," comments anthropologist Meredith Small.[164] "We might think that because we have culture—and with it all kinds of medical interventions and technologies—that we are immune from natural selection, but nature proceeds as usual." In the developing world, infectious diseases still cut a deadly swathe through infant populations. At least 1,200 children under the age of five die of malaria every day. The mounting pace of global travel and

(Opposite) Only a handful of hunter-gatherer populations still follow a traditional mobile existence. Anthropologist and Leakey grantee Steve Lansing studies the Cave Punan people of Indonesian Borneo, who communicate with each other using a unique song language. Their DNA shows them to be direct descendants of Ice Age hunter-gatherers.

migration means that even developed nations are vulnerable, as the worldwide coronavirus epidemic grimly reminded us. Since pathogens constantly evolve to subvert our immune defenses, our battle against new epidemics will never be over.

Meanwhile, the specter of catastrophic climate change looms over our long-term future. The technological forces that contributed to our adaptability as a species have set loose forces that may soon render the Earth unlivable for us. As we all face this alarming prospect, the science of humanity's origins can serve as an important touchstone. Richard Leakey believes that "the link between the past and the future, as seen in the present, is fundamental to stewardship of the climate."[165] Studying the past informs us of the scale and impacts of previous episodes of climate change on our ancestors and helps to define our place in the natural order. An evolutionary perspective reminds us of the contingent nature of our presence on Earth—the sheer luck and improbability behind our rise to dominance of the natural world—and of how extinction is the ultimate fate of every species.

This perspective might seem depressing, yet ever since Darwin wrote of the "grandeur in this view of life" in the final sentence of *On the Origin of Species*, others have echoed his sentiment that the science of origins should instill us with wonder and an appreciation of our role in the story of life. Biologist Kenneth Miller comments that "we may have started as just one branch on

Darwin's tangled riverside bank of life, but we are the branch that emerged to make sense of it all...It is important, vitally important, that we get the details of that journey right, that we tell the truth about our biological origins, and that we draw the wisdom of experience from an understanding of our evolutionary past."[166]

From Louis and Mary Leakey's trailblazing fossil digs at Olduvai Gorge to today's gleaming arrays of DNA sequencing machines, the past half-century of discovery has transformed our understanding of our past. The last decade alone has seen exponential gains in our ability to explore human genomes and retrieve intimate biological clues from tiny scraps of ancient

(Above) In Uganda, field assistant James Kyomuhendo collects data as part of the Kibale Chimpanzee Project. The Kibale scientists are involved with a local community organization, the Kisiisi Water Project, which supports education, health, and care for the environment in local schools (opposite). Outreach to local African communities is crucial to the survival of today's rapidly dwindling populations of primates.

human bones. With advanced imaging and mapping data, we are able to identify and record more archaeological sites than ever before, and extract data with undreamed-of precision. In Africa, where the most important fossil discoveries were made by outsiders from elite universities, indigenous scholars are finally emerging to lead the next generation in understanding their nation's heritage.

But in the midst of all these positive advances, one area—the study of primates—is seriously losing ground. In a comprehensive survey published in 2017, it was estimated that about 75 percent of the world's 500 or more primate species are in decline, and some 60 percent are directly threatened with extinction. The pressures of a rapidly warming climate, expanding farmland, logging, mining, and illegal hunting are chiefly to blame. Apart from the unique insights that primates offer into human evolution and behavior, they often play the role of a keystone species, essential for maintaining the health and biodiversity of tropical forests.

Orangutans are among the most endangered populations. In Borneo, more than half their population has perished in the last decade and a half. Even as these graceful creatures are pushed to the brink, a dramatic new discovery of a previously unsuspected species, the Tapanuli orangutan, was made in 2017 by genetic researchers supported by The Leakey Foundation. With just 800 survivors in the remote upland forests of Sumatra, the Tapanuli is probably the rarest of all the great apes, and their small population may soon lose out to mining activity and a projected dam. Without quick intervention, the researchers predicted, "we may see the discovery and extinction of a great ape species within our lifetime."[167]

Human pressures such as deforestation and the bushmeat trade are taking a steep toll of chimpanzee populations, along with their special cultural traditions, such as nut-cracking, or probing for ants or termites with sticks. A recent Africa-wide survey of chimpanzee communities showed that wherever chimp populations are suffering disturbance, such cultural practices are less likely to be performed. The researchers blame this on the increasing loss of older animals, who are instrumental in passing down these learned behaviors to the next generation. Without such knowledge, the young are deprived of important high-energy foods, while science loses insights into the inventiveness and adaptability of our closest primate cousins.

The outlook for primates is not entirely bleak. A few striking success stories clearly demonstrate how outreach and cooperation with local communities, coupled with fundraising and ecotourism, can help individual populations thrive. The poster child for such efforts is the Dian Fossey Gorilla Fund's Karisoke Research Center in central Africa. According to one estimate in 1981, there were just 254 mountain gorillas remaining there, and Dian Fossey was pessimistic about their future. Despite ongoing violence in the Democratic Republic of Congo, the last census in 2018 reported over 600 mountain gorillas in the Virunga Mountains, with a total population in Africa of over 1,000. Meanwhile, in the late 1970s, Brazil's golden lion tamarins were thought to have dwindled to only around 70, and their prospects looked bleak. Now, as the result of captive breeding at the Smithsonian's National Zoo and extensive publicity in Brazil, there are thought to be more than 3,000. Time may be running out for many species, but the broad recipe for sustainable conservation efforts is clear.

Humanity's impact on wildlife and on Earth's resources was a constant concern of Louis Leakey in his final years. At a public lecture in 1971, Leakey affirmed that "I believe in the future of man...but time is getting short. We stand, today, at a very important crossroads. We must choose what may or may not be the next milestone in man's twenty-million-year journey. It is the young who have to see the future and, with our generation's help, use the powers of conceptual thought and reason to ensure that our delightful heritage will not be lost."[168]

(Opposite) Louis Leakey in 1966, holding the tooth of *Deinotherium*, an extinct relative of today's elephants. "People frequently ask me why I devote so much time to seeking out facts about man's past," Leakey said. "The past shows clearly that we all have a common origin and that our differences in race, color and creed are only superficial."

ACKNOWLEDGMENTS

For fifty years, The Leakey Foundation has helped support the research behind many of the extraordinary advances in our understanding of human origins and the biology and behavior of primates. The idea of celebrating that achievement with a book of essays on fifty groundbreaking discoveries, first suggested by Melissa Jensen, was developed and led by the Foundation's president, Camilla Smith, whose enthusiasm and shrewd judgments were a constant source of inspiration and common sense throughout the project.

Paula Apsell, Executive Producer Emerita of NOVA, offered me the opportunity and encouragement to write the book while serving as NOVA's Senior Science Editor. I am grateful for many years of fruitful collaboration with Paula on NOVA programs and series on human evolution such as *In Search of Human Origins* and *Becoming Human*. Continuing support was also provided by John Bredar, Vice President for National Programming at WGBH, and my NOVA colleagues, including Co-Executive Producers Julia Cort and Chris Schmidt. Among the distinguished scientists I've been fortunate to work with on NOVA projects, I'd like to single out Kirk Johnson, Sant Director of the National Museum of Natural History and host of the NOVA series *Making North America* and *Polar Extremes*, who contributed the Foreword, and paleoanthropologist Don Johanson, a source of guidance and friendship since he hosted NOVA's *In Search of Human Origins* series in 1994. I've also benefitted from many shared ideas and discussions with independent producers Graham Townsley, Larry Klein, and Niobe Thompson.

Discovering Us was a huge team effort led by Camilla and her dedicated staff at The Leakey Foundation, notably Sharal Camisa and Meredith Johnson. Meredith writes and produces The Leakey Foundation's award-winning podcast series *Origin Stories*, several episodes of which formed the inspiration for essays in the book. Other book team members who helped make the book a reality were Andrew Smith, Hilary McClellen, and Jeanne Newman.

Members of The Leakey Foundation's Science Executive Committee (SEC) provided expert help with research and advice in steering the project, notably John Fleagle, Kristen Hawkes, Nina Jablonski, Steve Kuhn, John Mitani, Robert Seyfarth, Joan Silk, and Anne Stone. I owe a special debt to Dan Lieberman, who made himself constantly available as a source of up-to-date perspective and guidance.

For assistance and reviews of the material in this book, I'm grateful to the following: Rebecca Ackerman, Susan Alberts, Mark Aldenderfer, Zeray Alemseged, Bridget Alex, Monty Alexander, Chris Beard, Christophe Boesch, George Chaplin, Steve Churchill, Liesl Clark, Nicholas Conard, Deborah Cunningham, Iain Davidson, Jeremy DeSilva, Anna Di Rienzo, Marina Elliott, Dean Falk, Eduardo Fernandez-Duque, David Frayer, Ann Gibbons, Laurie Godfrey, Jessica Hartel, John Hawks, John Hoffecker, Sarah Hrdy, Bill Jungers, Bill Kimbel, Cheryl Knott, María Martinón-Torres, Sharon McFarlin, Virginia Morell, Liza Moscovice, Sherry Nelson, Isaiah Nengo, Susan Perry, Rick Potts, Jill Pruetz, Anne Pusey, Alessia Ranciaro, Kaye Reed,

Philip Rightmire, Michael Rogers, Stacy Rosenbaum, Sileshi Semaw, Chalachew Seyoum, John Shea, Sarah Tishkoff, Peter Ungar, Tim White, Bernard Wood, and Brian Wood. I would particularly like to thank Eugene Harris and John Langdon for their help with the stories on ancient genetics.

Finally, my deepest gratitude of all goes to my wife, Janet. Her shrewd editorial comments and vital support helped keep this old fossil sane during many months of research and writing.

SOURCES OF QUOTATIONS

INTRODUCTION

1. Morell, V. *Ancestral Passions*, Simon & Schuster, 1995, p. 181.
2. Ibid., p. 194.
3. Gould, S. J. *Wonderful Life,* Norton, 1989, p. 31.
4. Tattersall, I. *Masters of the Planet,* St Martin's Press, 2012, p. 87.
5. Johanson, D., & Edgar, B. *From Lucy to Language*, Simon & Schuster, 2006, p. 33.
6. King, T. "Sex, Power and Ancient DNA," *Nature*, v. 555, March 13 2018, p. 307–308.
7. Hawks, J. "Human Evolution Is More a Muddy Delta than a Branching Stream," *Aeon* website, February 8 2016.
8. Lewin, R. "The Old Man of Olduvai Gorge," *Smithsonian* website, October 2002.
9. Betts, C. A. "Man Evolved Like Animals," *Science Newsletter*, April 17 1965, p. 243.
10. Lewin, "Old Man of Olduvai Gorge."
11. Morell, *Ancestral Passions*, p. 363.
12. Ibid., p. 237.
13. Stanford, C. *The New Chimpanzee*, Harvard University Press, 2018, p. 195.
14. Morell, *Ancestral Passions*, p. 238.
15. McKie, R. "Jane Goodall: 50 years...," *The Guardian*, June 26 2010.
16. Interview in The Leakey Foundation Oral History Project, Bancroft Library, UC Berkeley, 2003–2004, p. 18.
17. Morell, V. "Called 'Trimates'..." *Science,* April 16 1993, v. 260, 5106, p. 420–425.
18. Interview in Leakey Foundation Oral History Project, p. 20.
19. Stanford, *New Chimpanzee*, p. 9–10.
20. Quammen, D. "Fifty Years at Gombe," *National Geographic,* October 2010.
21. Ibid.
22. McKie, R. "Chimps with Everything," *The Guardian*, June 26 2010.
23. Morell, "Called 'Trimates.'"
24. Ibid.
25. Ibid.
26. Interview in Leakey Foundation Oral History Project, p. 5.
27. Galdikas, B. *Reflections of Eden,* Little Brown, 1995, p. viii.
28. Ibid., p. 203.
29. Morell, "Called 'Trimates.'"
30. Cole, S. *Leakey's Luck*, Harcourt Brace, 1975, p. 402.
31. Interview in Leakey Foundation Oral History Project, p. 22.
32. Morell, *Ancestral Passions*, p. 400.
33. Clark, J. D. "Mary Douglas Leakey," *Proceedings of the British Academy,* 2001, 111, p. 611–612.
34. Morell, *Ancestral Passions*, p. 203, footnote.
35. Leakey, M. *Disclosing the Past*, Doubleday, 1984, p. 177.
36. Clark, "Mary Douglas Leakey," p. 613.
37. Leakey, R. "Five Questions for Richard Leakey," YouTube interview posted by Stony Brook University, October 27 2016.
38. Morell, *Ancestral Passions*, p. 158.
39. Smith, B. "First Family of Paleoanthropology," *Stony Brook University Magazine*, Issue 1, 2016, p. 20.
40. Ibid.
41. Perry, S. "The Importance of Long-term Research Sites," *AnthroQuest*, The Leakey Foundation Newsletter, 33, 2016, p. 12.
42. Mitani, J. "Primate Research Fund provides a lifeline...," The Leakey Foundation press release, October 7 2020.

DISCOVERING US

1. "Discovery of Oldest Primate Skeleton...," American Museum of Natural History news release, June 5 2013.
2. Wilford, J. N. "Palm-Size Fossil Resets Primates' Clock...," *New York Times*, June 5 2013.
3. Nengo, I. "Discovering Alesi: A Timeline," *AnthroQuest,* The Leakey Foundation Newsletter, 36, 2017, p. 3.
4. "Ancestor," *Origin Stories*, The Leakey Foundation podcast, season two, episode 1, August 31 2017.
5. Ibid.
6. Nengo, I. "Great Ape Haters," *Anthropology News,* American Anthropological Association, September 18 2018.
7. Gibbons, A. "Anthropologists Bet on Their Latest Data...," *Science*, v. 256, 5055, April 17 1992, p. 308–309.
8. Ibid., box 2.
9. Jungers, B. Personal comment to author.
10. Goodall, J. *Through a Window*, 1990, Soko Publications, p. 108.
11. "How Infighting Turns Toxic for Chimpanzees," The Leakey Foundation blog, March 28 2018.
12. "Ngogo Chimpanzees on Patrol," The Leakey Foundation press release, June 23 2017.
13. Goodall, *Through a Window*, p. 108–109.
14. "An Interview with Frans de Waal," *Voices for Biodiversity* website, November 2017.
15. Hartel, J. Interview with the author.
16. Ibid.
17. Gibbons, A. *The First Human,* Doubleday, 2006, p. 3–4.
18. Djimdoumalbaye, A. "La Découverte par Ahounta Djimdoumalbaye," Toumaï, Paleotchad website, undated.
19. Tattersall, I. *Masters of the Planet*, St Martin's Press, 2012, p. 7.
20. Marshall, J. "Who'd Make a Monkey...," *Times Higher Educational Supplement,* October 25 2002, p. 19.
21. Gibbons, A. "A New Kind of Ancestor: *Ardipithecus* Unveiled," *Science,* v. 326, October 2 2009, p. 40.
22. Ibid.
23. *Discovering Ardi*, Discovery Channel documentary, 2011.

24. DeSilva, J. "Where Do We Begin?," *Natural History*, v. 126, 8, September 2018, p. 12.

25. Mosendz, P. "Lucy's 40th birthday...," *Newsweek*, November 30 2014.

26. Lieberman, D. E. *The Story of the Human Body*, Vintage Books, 2013, p. 45.

27. "Anthropologist Donald Johanson on 'Lucy's Legacy'," *Science Friday*, National Public Radio, March 6 2009.

28. Ward. C. "Australopiths," in J. Gurche, ed., *Lost Anatomies*, Abrams, 2019, p. 63.

29. Greshko, M. "Newfound Footprints Stir Debate," *National Geographic* website, December 14 2016.

30. Hatala, G. et al. "Laetoli Footprints Reveal a Bipedal Gait...," *Proceedings of the Royal Society B*, August 3 2016.

31. Johanson, D., & Edgar, B. *From Lucy to Language*, Simon & Schuster, 2006, p. 142.

32. NOVA, "Becoming Human," PBS website, episode 1 transcript, November 3 2009.

33. Maxmen, A. "Digging through the World's Earliest Graveyard," *Nautilus*, September 25 2014.

34. Ibid.

35. Ungar, P. S. "The Real Paleo Diet," *Scientific American*, July 2018, p. 46.

36. Keyser, A.W. "Finds in South Africa," *National Geographic* magazine, May 2000, p. 78.

37. Ungar, P. Interview with author.

38. Semaw, S. Personal communication with author.

39. Rogers, M. Interview with author.

40. Ibid.

41. Ibid.

42. Boesch, C., & Boesch-Achermann, H. "Dim Forest, Bright Chimps," *Natural History*, September 1991, p. 76.

43. Pruetz, J. Personal communication with author.

44. De Waal, F. *The Bonobo and the Atheist*, Norton, 2013, p. 69.

45. Ibid., p. 70.

46. Moscovice, L. Interview with author.

47. Ibid.

48. Angier, N. "In the Bonobo World, Female Camaraderie Prevails," *New York Times*, September 10 2016.

49. Moscovice, interview with author.

50. De Waal, *Bonobo*, p. 192, footnote 2.

51. Ibid.

52. Seyoum, C. M. "Leakey Grantee Finds Oldest Jawbone of Genus *Homo*," *AnthroQuest*, The Leakey Foundation Newsletter, 32, 2015.

53. Ibid.

54. Kimbel, W. "Origins of Genus *Homo*," CARTA lecture, UCTV online, April 20 2016.

55. Seyoum, "Leakey Grantee Finds Oldest Jawbone."

56. "Discovery of 2.8-Million-Year-Old Jaw...," *Science Daily*, March 4 2015.

57. Kimbel, W., & Villmoare, B. "From *Australopithecus* to *Homo*: The Transition that Wasn't," *Philosophical Transactions of the Royal Society B*, v. 371, 20150248, 2016.

58. Reed, K. Personal communication with author.

59. Leakey, R. E., & Lewin, R. *Origins*, Macdonald and Jane's, 1977, p. 86.

60. Leakey, R. *One Life*, Salem House, 1984, p. 149.

61. Koobi Fora Field Journal, September 9 1972, quoted in Morell, V. *Ancestral Passions,* Simon & Schuster, 1995, p. 399.

62. Leakey & Lewin, *Origins*, p. 86.

63. Ibid.

64. Wood, B. "Fifty Years after *Homo habilis*," *Nature*, v. 508, April 3 2014, p. 33.

65. Tattersall, *Masters of the Planet*, p. 122.

66. Lorch, D. "Old Hand at Finding the Oldest Hominids," *New York Times*, November 7 1995.

67. NOVA, Becoming Human, PBS website, episode 2 transcript, November 8 2009.

68. Dreifus, C. "Born, and Evolved, to Run," *New York Times*, August 22 2011.

69. Chen, I. "Born to Run," *Discover Magazine*, May 28 2006.

70. Cohen, J. *Almost Chimpanzee*, Times Books, 2010, p. 216.

71. Gorman, R. M. "Cooking Up Bigger Brains," *Scientific American* website, January 1 2008.

72. Jha, A. "Scientists Find Clue to Human Evolution's Burning Question," *The Guardian*, April 2 2012.

73. Rosenbaum, S. Interview with author.

74. Ibid.

75. Ibid.

76. Lents, N., & Rosenbaum, S. "What Mountain Gorillas Can Teach Us about Gendered Behaviors," *Human Evolution* blog, June 6 2016.

77. McFarlin, S. Interview with author.

78. "Grantee Spotlight: Shannon McFarlin," The Leakey Foundation website, November 20 2015.

79. DiConsiglio, J. "Gorilla Graveyard Yields Science Secrets," George Washington University press release, December 10 2014.

80. Vogel, E. Interview with author.

81. Ibid.

82. Dell'Amore, C. "Orangutans Are More Like Us than You Think," *National Geographic News,* April 21 2016.

83. Vogel, interview with author.

84. Ibid.

85. Conard et al. "Excavations at Schöningen and Paradigm Shifts in Human Evolution," *Journal of Human Evolution*, 89, 2015, p. 4.

86. Balter, M. "The Killing Ground," *Science*, v. 344, 6188, June 6 2014, p. 1082.

87. Tierney, J. "The Search for Adam and Eve," *Newsweek*, v. 111, January 11 1988, p. 46–52.

88. Wilkins, A. "The Scientists behind Mitochondrial Eve...," *io9* blog, January 27 2012.

89. Relethford, J. *Reflections of Our Past*, Routledge, 2018, p. 53.

90. Ibid.

91. Zimmer, C. "Oldest Fossils of Homo Sapiens..." *New York Times,* June 7 2017.

92. "Rewriting Our Story," *Origin Stories*, The Leakey Foundation podcast, episode 29.

93. Greshko, M. "These Early Modern Humans...," *National Geographic* website, June 7 2017.

94. Ibid.

95. Wong, K. "Ancient Fossils from Morocco...," *Scientific American* website, June 8 2017.

96. Hawks, J. "Three Big Insights...," *Medium* website, January 3 2019.

97. "Our Fractured African Roots," press release, Max Planck Institute, July 12 2018.

98. Hays, B. "Humans evolved in small groups...," *Science News*, July 11 2018.

99. Leakey, M. *Disclosing the Past*, Doubleday, 1984, p. 83.

100. "Evolution of Human Innovation," National Museum of Natural History website.

101. Potts, R. Interview with author.

102. Public lecture to Leakey Foundation, May 2 2018, recorded by author.

103. King, W. "The Reputed Fossil Man of the Neanderthal," *Quarterly Journal of Science*, 1864, p. 88–97.

104. Tattersall, *Masters of the Planet*, p. 177.

105. Scott, J. "Smarter Than You Think," *Coloradan*, December 1 2016.

106. Sample, I. "Neanderthal Genes Increase Risk of Serious COVID-19, Study Claims," *The Guardian*, September 9 2020.

107. Reich, D. *Who We Are and How We Got Here*, Pantheon, 2018, p. 50.

108. Shreeve, J. "Case of the Missing Ancestor," *National Geographic*, July 2013.

109. Hawks, J. "Ancient Genomes 1: The Denisovans," YouTube lecture, June 24 2014.

110. Sample, I. "Offspring of Neanderthal and Denisovan...," *The Guardian* website, August 22 2018.

111. Harris, K., & Nielsen, R. "Q&A: Where did the Neanderthals Go?," *BMC Biology*, v. 15, 2017, p. 73.

112. Ibid.

113. Cook, J. "The Lion Man...," British Museum blog, October 10 2017.

114. Lobell, J. "New Life for the Lion Man," *Archaeology Magazine*, v. 65, March/April 2012, p. 2.

115. Chauvet, J-M. et al. *Chauvet Cave: The Discovery of the World's Oldest Paintings*, Thames & Hudson, 1996, p. 41–42.

116. "The Grandmother Hypothesis," *Origin Stories*, The Leakey Foundation podcast, episode 15, transcript, p. 3.

117. "Why Grandmothers May Hold the Key...," *Goats & Soda*, National Public Radio, interview, June 7 2018.

118. Hrdy, S. "Comes the Child...," *Evolutionary Anthropology*, 21, p. 188.

119. Callier, V. "For Baboons, a Tough Childhood Can Lead to a Shorter Life," *Smithsonian* website, April 19 2016.

120. Alberts, S. Interview with author.

121. Ibid.

122. Alberts, S. C. "The Challenge of Survival for Wild Infant Baboons," *American Scientist*, v. 104, 2016, p. 373.

123. Wade, N. "How Baboons Think (Yes, Think)," *New York Times*, October 9 2007.

124. Cheney, D. L. & Seyfarth, R. M., *Baboon Metaphysics*, University of Chicago Press, 2007, p. 97.

125. Wade, N. "Chimps and Monkeys could talk. Why don't they?," *New York Time*, January 12 2010.

126. "Rising Star," *Origin Stories*, The Leakey Foundation podcast, episode 26.

127. Ibid.

128. Ibid.

129. Barras, C. "Meet Neo, the Most Complete Skeleton...," *New Scientist*, v. 3125, May 13 2017.

130. NOVA, "The Violence Paradox," PBS website, transcript, November 20 2019.

131. Handwerk, B. "An Ancient, Brutal Massacre...," *Smithsonian* website, January 20 2016.

132. Ibid.

133. Wrangham, R. *Demonic Males,* Houghton Mifflin Harcourt, 1996, p 63.

134. Pinker, S. *The Better Angels of Our Nature*, Viking, 2011, p. xxiv.

135. Morgan, J. *The Life and Adventures of William Buckley,* Australian National University Press, 1965–1991, p. 71–72.

136. Jablonski, N. G. *Skin: A Natural History*, University of California Press, 2006, p. 65.

137. Gibbons, A. "Shedding Light on Skin Color," *Science*, v. 346, 6212, November 21 2014, p. 934–936.

138. Gibbons, A. "New gene variants reveal the evolution of human skin color," *Science* website, October 12 2017.

139. Riley, A. "How Tibetans Survive Life on the 'Roof of the World'...," *BBC Future* website, February 27 2017.

140. Bigham, A. Interview with author.

141. Zimmer, C. "In Ancient Skeletons, Scientists Discover a Modern Foe...," *New York Times*, May 9 2018.

142. "Penn Team Links Africans' Ability to Digest Milk...," *Penn Today* press release, March 13 2014.

143. Swaminathan, N. "African Adaptation to Digesting Milk...," *Scientific American* website, December 11 2006.

144. Renaud, J-P. "Monkey Business," *UCLA Magazine*, July 1 2015.

145. Boser, U. M. "Why Trust Matters," Ulrich Boser blog, June 8 2014.

146. "Capuchins Traditions Project," Lomas Barbudal Monkey Project, UCLA Dept. of Anthropology website, undated.

147. Boser, "Why Trust Matters."

148. Ibid.

149. De Waal, F. *The Age of Empathy*, Crown, 2009, p. 165.

150. Zimmer, C. "First, Kill the Babies," *Discover Magazine*, September 1 1996.

151. Sridhar, H. "Revisiting Hrdy 1974," *Reflections on Papers Past*, Hari Sridhar blog, January 2 2018.

152. Packer, C. "Infanticide Is No Fallacy," *American Anthropologist*, v. 102, 4, 2001, p. 829–857.

153. Ebersole, R. "Monkeys from Heaven," *National Wildlife*, June 1 2003.

154. Concar, D. "Inhuman Futures," *New Scientist*, December 11 1999, p. 2216.

155. Fernandez-Duque, E. Interview with author.

156. Ibid.

157. Ibid.

158. Dell'Amore, C. "Owl Monkeys Shed Light...," *National Geographic* website, February 14 2013.

159. Cordain, L. *The Paleo Diet*, Wiley, 2001, p. 14–15.

160. McKie, R. "Is Human Evolution Finally Over?," *The Guardian,* February 3 2002.

161. Gould, S. J. "The Spice of Life," *Leader to Leader*, Frances Hesselbein Leadership Forum, University of Pittsburgh, Winter 2000, p. 19.

162. Keim, B. "Human Evolution Enters an Exciting New Phase," *Wired* website, November 29 2012.

163. Ward, P. "What May Become of *Homo Sapiens*," *Scientific American*, December 2012.

164. Small, M. "Are Humans Still Evolving?," *Scientific American* website, October 21 1999.

165. Dwyer, J. "Warning of a World that's Hotter and Wetter," *New York Times*, May 22 2012.

166. Miller, K. R. *The Human Instinct*, Simon & Schuster, 2018, p. 230.

167. "New Species of Orangutan Announced," World Wildlife Fund press release, November 2 2017.

168. Leakey, L. S. B. "Excerpt from a Speech by Dr Louis S. B. Leakey," The Leakey Foundation, fundraising leaflet, 1971.

GLOSSARY

Adaptation – a process of change in which a plant or organism becomes better suited to its environment.

Allele – one of the two or more variations of a specific gene caused by mutation.

Archaic human – an individual belonging to one of several species of human ancestors that are distinct from anatomically modern humans.

Aurignacian – an archaeological culture or tradition linked to the first major arrival of modern humans in Europe around 43,000 years ago.

Australopithecine or **Australopith** – a member of several extinct species of hominins that lived in East and South Africa from 4 to 2 million years ago. They were closely related to, if not direct ancestors of, the earliest *Homo* lineage.

Australopithecus – a specific genus or group of extinct australopithecines.

Base pairs – two chemical bases bonded together to form a "rung" of the DNA double helix "ladder."

Chromosome – a thread-like structure of DNA sequences contained in the cells of plants and animals.

Denisovan – an extinct species of archaic human discovered by DNA analysis in 2010. At the time of publication, no complete enough fossil has been found and confirmed, so a new species name has not yet been assigned to it. The Denisovans were a widespread population in Eurasia roughly contemporary with the Neanderthals.

DNA – or deoxyribonucleic acid is a self-replicating material which is present in nearly all living organisms as the main constituent of chromosomes. It is the carrier of genetic information.

Epigenetic – biological effects caused by chemical processes that change the expression of genes (i.e., by switching them on or off) rather than by altering the DNA sequence itself.

Genetic drift – one of the basic mechanisms of evolution, involving gradual random change in the genetic makeup of a population as some individuals leave behind more descendants than others.

Genome – an organism's complete set of genetic instructions, encoded in DNA.

Genus – a group of plants or animals that share many characteristics; more closely related than a family but less similar than a species.

Haplotype – a distinctive genetic "signature," consisting of a set of DNA variations, or polymorphisms, that tend to be inherited together. A haplotype can refer either to a combination of alleles or to a set of single nucleotide polymorphisms, or SNPs, on the same chromosome.

Hominid – a member of the biological family known as Hominidae, comprising all modern and extinct Great Apes, including humans.

Hominin – a member of the "tribe" of primates known as Homininae, of which only one species exists today: *Homo sapiens*. The term also applies to extinct species that once belonged to the *Homo* lineage.

Homo erectus – an extinct ancient human ancestor, or "archaic human," first appearing in the fossil record about 2 million years ago. Its origins lay in east Africa or possibly in Eurasia, from where it is thought to have spread rapidly to the rest of Asia and Southeast Asia.

Homo floresiensis – an extinct ancestor so far known only from the island of Flores, Indonesia. Dating to between about 100,000–60,000 years ago, this hominin's tiny stature and brain, and other distinctive features, earned it a nickname, the "Hobbit."

Homo habilis – one of the first members of the genus *Homo*, known from sites in east Africa dated to roughly 2.4–1.4 million years ago. Despite the designation of the fossil as "handy man," it is uncertain if early *Homo* or the australopithecines were the first to regularly use stone tools.

Homo heidelbergensis – a ruggedly built, archaic human ancestor known from Africa and possibly Asia from about 700,000 to 200,000 years ago. It is also thought to have been the first to adapt to the cold conditions prevailing in Europe's higher latitudes.

Homo neanderthalensis – our closest extinct human relative, occupying a wide area of Europe and central and southwest Asia between about 400,000 to 40,000 years ago. They were robustly built to help withstand the rigors of Ice Age climate extremes.

Lactase – an enzyme that breaks down the milk sugar lactose, facilitating the digestion of milk and other dairy products.

Mitochondria – the powerhouses that convert chemical energy from food inside a cell. They consist of an organelle or cell structure enclosed by a double membrane, which contains a small circular chromosome with 37 genes. Mitochondria and their DNA are passed down exclusively by mothers to their offspring.

Modern humans – a term applied to fossils of ancient humans that share many anatomical features with today's populations.

Molecular clock – a technique used to time evolutionary events by measuring the number of changes, or mutations, in the gene sequences of species over time.

MtDNA – see **Mitochondria**

Mutation – a change in a DNA sequence that occurs either because of random "copying mistakes" during cell division, or because of exposure to environmental factors that can damage DNA, such as cigarette smoke.

Natural selection – a major driver of evolution, and a natural process that results in the survival and reproductive success of those organisms best adapted to their environments.

Neanderthal – see *Homo neanderthalensis*

Nuclear DNA – the DNA contained in the nucleus of animal cells, inherited through both parents.

Nucleotide – one of the structural building blocks of DNA and RNA molecules. They consist of a base, one of four chemicals abbreviated to A, C, G, and T, plus a molecule of sugar and one of phosphoric acid.

Obsidian – a naturally occurring, glassy volcanic rock that can be worked into blades much sharper than the finest steel.

Paleoanthropologist – a specialist in the study of the origin and development of humanity, largely through its fossil remains.

Paleolithic – referring to the earliest period of the Stone Age, spanning over 2 million years during the late Pliocene and Pleistocene epochs.

Paranthropus – a genus of australopithecines, including the species *Paranthropus boisei*, famous for its first discovery by Mary Leakey in Olduvai Gorge in 1959, and *Paranthropus robustus*, found by Robert Broom in South Africa in 1938. Nicknamed "Nutcracker Man."

Recombination – the exchange of DNA between parents during reproduction, which results in novel combinations of traits in their offspring.

Sexual dimorphism – a marked difference in physical characteristics between the males and females of any species, usually referring to distinctive differences in size.

Single nucleotide polymorphism (or SNP) – a particular DNA sequence where the nucleotide bases (the A, C, G, or T "letters" of the genetic code) are different, either between two different people or between paired chromosomes.

SELECTED SOURCES AND SUGGESTED READING

The following list is a limited, personal selection of titles that the author found most useful in writing this book.

GENERAL SURVEYS OF HUMAN ORIGINS

Harcourt, A. H. *Humankind: How Biology and Geography Shape Human Diversity*, Pegasus Books, 2015. A wide-ranging exploration of the biological and cultural diversity of populations around the globe and its evolutionary origins.

Humphrey, L., & Stringer, C. *Our Human Story*, Natural History Museum, London, 2018. Brief, well-illustrated general guide to human origins.

Johanson, D., & Edgar, B. *From Lucy to Language*, 2nd edition, Simon & Schuster, 2006. A lavishly illustrated catalog of major ancient hominin fossils.

Lee, S-H. *Close Encounters with Humankind*, Norton, 2015. Engaging, easy-to-read short popular essays on major mysteries in human evolution and how they affect our present-day lives and behavior.

Lieberman, D. *The Story of the Human Body*, Vintage Books, 2013. A highly accessible exploration of human evolution and how it continues to affect our bodies and health today.

Stringer, C., & Andrews, P. *The Complete World of Human Evolution*, 2nd edition, Thames & Hudson, 2012. A well-illustrated overview of primate and hominin fossil origins.

Tattersall, I. *Masters of the Planet,* St. Martin's Press, 2012. An engrossing account of fossil hominin discoveries.

Wood, B. *Human Evolution: A Very Short Introduction,* Oxford University Press, 2019. A pocket-sized survey of major ancestral finds and their broader significance.

SPECIFIC INSIGHTS AND CONTROVERSIES ON HUMAN ORIGINS

Bahn, P. G. *Images of the Ice Age*, Oxford University Press, 2016. Latest edition of the classic illustrated survey of Ice Age cave art.

Falk, D. *The Fossil Chronicles*, University of California Press, 2011. The story of the heated controversies surrounding the discovery of the "Hobbit" of Flores, Indonesia.

Gurche, J. *Lost Anatomies*, Abrams, 2019. A striking portfolio of artistic renderings and reconstructions of ancient human fossils, based on expert anatomical knowledge.

Jablonski, N. G. *Skin: A Natural History*, University of California Press, 2006. A fascinating and authoritative guide to evolutionary insights on skin function, color, and racist theories.

Johanson, D. C., & Wong, K. *Lucy's Legacy*, Harmony Books, 2009. The discoverer of the celebrated "Lucy" fossil looks back on a lifetime of discovery and on exciting new developments in human origins.

Langdon, J. H. *The Science of Human Evolution*, Springer, 2016. Case studies of controversies in the study of human origins.

Lorblanchet, M., & Bahn, P. *The First Artists*, Thames & Hudson, 2017. New evidence for the world's oldest art and symbolism by two leading authorities.

MacPhee, R. D. E. *End of the Megafauna*, Norton, 2019. A vividly illustrated exploration of the role of humans and climate in the extinction of Ice Age beasts and survivors such as the Giant Sloth Lemurs of Madagascar.

Morwood, M., & Oosterzee, P. V. *A New Human*, Smithsonian Books, 2007. The amazing story of the fossil "Hobbit" of Flores, Indonesia, by its lead discoverer, the late Mike Morwood.

Papagianni, D., & Morse, M.A. *The Neanderthals Rediscovered*, Thames & Hudson 2013. Concise, well-illustrated guide to the vanished Neanderthals.

Petzinger, G. V. *The First Signs: Unlocking the Mysteries of the World's Oldest Symbols*, Simon & Schuster, 2016. A lively and personal view of the earliest evidence for symbolism in prehistoric rock art.

Reader, J. *Missing Links: In Search of Human Origins*, Oxford University Press, 2011. A British writer and photojournalist's account of his lifetime of fascination with fossil human discoveries, notable for his artful photos.

Tattersall, I. *The Strange Case of the Rickety Cossack,* St. Martin's Press, 2015. An engaging view of changing currents in the study of human origins.

Ungar, P. S. *Evolution's Bite: A Story of Teeth, Diet, and Human Origins*, Princeton University Press, 2017. The surprising lessons from fossil dental evidence about our ancestors' diet, behavior, and environment.

Wrangham, R. *Catching Fire: How Cooking Made Us Human*, Basic Books, 2009. A leading anthropologist's influential theory about the impact of meat and cooking on human evolution.

Zuk, M. *Paleofantasy*, Norton, 2013. A provocative and frequently hilarious exposé of the fashionable craze for "Paleo" diets and what the evidence from prehistory actually suggests.

GENETICS AND HUMAN EVOLUTION

Harris, E. E. *Ancestors in Our Genome*, expanded paperback edition, Oxford University Press, 2019. A well-written introduction to current genetic techniques and discoveries in human evolution.

Pääbo, S. *Neanderthal Man: In Search of Lost Genomes*, Basic Books, 2014. The inside story of the decoding of the Neanderthal genome.

Reich, D. *Who We Are and How We Got There*, Pantheon, 2018. A groundbreaking guide to the fast-changing study of ancient DNA by one of the field's leading scientists.

Relethford, J.E., & Bolnick, D. A. *Reflections of Our Past,* 2nd edition, Routledge, 2018. How genetics has transformed the study of human origins.

Rutherford, A. *A Brief History of Everyone Who Ever Lived*, The Experiment, 2016. An entertaining popular account of the impact of advances in ancient genetics.

Zimmer, C. *She Has Her Mother's Laugh: The Powers, Perversions, and Potential of Heredity*, Dutton, 2018. A distinguished U.S. science journalist explores controversial aspects of genetics research, such as DNA ancestry testing.

PRIMATE EMERGENCE AND BEHAVIOR

Begun, D. R. *The Real Planet of the Apes: A New Story of Human Origins*, Princeton University Press, 2016. New evidence in Asia, Europe, and Africa for the earliest ancestors of the great apes.

Cheney, D., & Seyfarth, R. M. *Baboon Metaphysics*, University of Chicago Press, 2008. Striking observations of the social world of baboons.

Ciochon, R. L., & Nisbett, R. A., eds. *The Primate Anthology: Essays on Primate Behavior, Ecology, and Conservation*, Prentice-Hall, 1998. An entertaining collection of articles by primate experts originally published in *Natural History* magazine.

De Waal, F. *Chimpanzee Politics*, Johns Hopkins, 1982. The classic account of conflict and reconciliation among captive chimpanzees.

De Waal, F., & Lanting, F. *Bonobo: The Forgotten Ape*, University of California Press, 1998. A fascinating introduction to a neglected primate cousin, with magnificent photographs.

Eckhart, G., & Lanjouw, A. *Mountain Gorillas*, Johns Hopkins, 2008. A lavishly illustrated and informative book on the lifeways and conservation of mountain gorillas.

Fossey, D. *Gorillas in the Mist*, Houghton Mifflin, 1983. The heroic and tragic story of Dian Fossey's life among mountain gorillas.

Galdikas, B. *Reflections of Eden: My Years with the Orangutans of Borneo,* Little Brown, 1995. A memoir by the leading orangutan primatologist.

Goodall, J. G. *Through a Window: My Thirty Years with the Chimpanzees of Gombe*, Houghton Mifflin, 1990. Jane Goodall's saga of her life and career among wild chimpanzees.

Hrdy, S. *Mothers and Others: The Evolutionary Origins of Mutual Understanding*, Harvard University Belknap Press, 2011. On the role of infanticide, female choice, and many other issues in human evolution.

Hrdy, S. *Mother Nature*, Pantheon, 1999. The relevance of anthropology and primate studies to understanding gender roles, mate choice, parenting, sex, and reproduction in the world today.

Perry, S. *Manipulative Monkeys: The Capuchins of Lomas Barbudal*, Harvard University Press, 2011. Groundbreaking insights into the highly complex social world of capuchins.

Sapolski, R. M. *A Primate's Memoir*, Scribner's, 2001. The story of a neuroscientist's two-decade life among the baboons of the Serengeti.

Stanford, C. *The New Chimpanzee*, Harvard University Press, 2018. An authoritative and highly readable account of the study of chimpanzee behavior.

IN-DEPTH TEXTBOOKS, PRIMARILY FOR STUDENTS

Ayala, F. J. & Cela-Conde, C. J. *Processes in Human Evolution*, Oxford University Press, 2017. Highly detailed survey of major human fossil discoveries.

Boyd, R. & Silk, J. B. *How Humans Evolved*, Norton, 1997, latest edition 2017. A comprehensive, well-illustrated student textbook on every aspect of human evolution.

De Waal, F. ed. *Tree of Origin: What Primate Behavior Can Tell Us about Human Social Evolution,* Harvard University Press, 2001. An important collection of essays by leading primatologists.

Fleagle, J. G. *Primate Adaption and Evolution*, Academic Press, 2013. A definitive survey of primate and hominin evolution.

Mitani, J. C., Call, J., Kappeler, M., Palombit, R. A., & Silk, J. B. *The Evolution of Primate Societies*, University of Chicago Press, 2012. A guide to the emergence and diversity of primate behavior.

Strier, K. B., ed. *Primate Ethnographies*, Pearson, 2014. A collection of autobiographical essays on the rewards and challenges of pursuing primate fieldwork.

INTRODUCTION: THE LEAKEY FAMILY AND CHANGING VIEWS OF HUMAN ORIGINS

Bowman-Kruhm, M. *The Leakeys*, Prometheus Books, 2010. A brief, highly readable account of the Leakey family saga.

Cole, S. *Leakey's Luck*, Harcourt, 1975. A classic biography of Louis Leakey, including the exploits of the "trimates" and the early story of The Leakey Foundation.

Gibbons, A. *The First Human*, Random House, 1981. A journalist's view of the quest for human origins in East Africa.

Leakey, M. *Disclosing the Past: An Autobiography*, McGraw Hill, 1986. Mary Leakey's autobiography.

Leakey, R. E. *One Life: An Autobiography,* Salem House, 1983. Absorbing personal account of the many achievements and challenges of Richard Leakey's life.

Leakey, R. E., & Lewin, R. *Origins,* E. P. Dutton, 1978. A popular account of Richard Leakey's fossil discoveries.

Lewin, R. "The Old Man of Olduvai Gorge," *Smithsonian* website, October 2002. A wide-ranging article on the life of Louis Leakey.

Morell, V. *Ancestral Passions*, Simon & Schuster, 1995. A compelling, in-depth biography of the Leakeys.

Morell, V. "Called 'Trimates'...," *Science,* April 16 1993, p. 260. On Jane Goodall, Biruté Galdikas, and Dian Fossey.

Quammen, D. "Fifty Years at Gombe," *National Geographic,* October 2010. On Jane Goodall.

FIFTY GREAT DISCOVERIES

1: PIONEERING PRIMATES

Beard, C. *The Hunt for the Dawn Monkey*, University of California Press, 2006.

Ni, X. et al. "The Oldest-Known Primate Skeleton," *Nature*, v. 498, 7452, June 6 2013, p. 60–64.

2: ALESI: BABY APE

"Ancestor," *Origin Stories*, The Leakey Foundation podcast, August 31 2017, season 2, episode 1.

Begun, D., & Gurche, J. "Planet of the Apes," *Scientific American* website, June 1 2006.

Benefit, B. "Evolution: Skull Secrets of an Ancient Ape," *Nature News*, v. 548, October 9 2017, p. 160–161.

Nengo, I. et al. "New Infant Cranium from the African Miocene Sheds Light on Ape Evolution," *Nature*, v. 548, 2017, p. 169–174.

3: MADAGASCAR'S LOST GIANTS

Gibbons, A. "Anthropologists Bet on Their Latest Data..." *Science*, v. 256, 5055, April 17 1992, p. 308–309.

Godfrey, L. R. "A New Interpretation of Madagascar's Megafaunal Decline," *Journal of Human Evolution*, v. 130, May 2019, p. 126–140.

Godfrey, L. R., & Jungers, W. L. "The Extinct Sloth Lemurs...," *Evolutionary Anthropology*, v. 12, 6, 2003, p. 252–263.

Goodman, S., & Jungers, W. L. *Extinct Madagascar*, University of Chicago Press, 2014.

4: THE BLOOD CLOCK

Langdon, J. H. *The Science of Human Evolution*, Springer, 2016, p. 44–46.

Pilbeam, D., & Lieberman, D. "Reconstructing the Last Common Ancestor of Chimpanzees and Humans," *Chimpanzees and Human Evolution*, Harvard University Press, 2017.

Relethford, J. *Reflections of Our Past,* Routledge, 2018, p. 33–34.

5: THE GREAT CHIMPANZEE WAR

Feldblum, J. T. et al. "The Timing and Causes of a Unique Chimpanzee Fission..." *American Journal of Physical Anthropology*, v. 166, 3, March 22 2018, p. 732–738.

Goodall, J. G. *Through a Window*, Houghton Mifflin, 1990, p. 108.

Langergraber, K. E. et al. "Group Augmentation, Collective Action...," *PNAS*, v. 114, 28, July 11 2017, p. 7337–7342.

Stanford, C. *The New Chimpanzee*, Harvard University Press, 2018.

6: CHIMPANZEE RECONCILIATION

De Waal, F. "Primates—a Natural Heritage of Conflict Resolution," *Science*, v. 289, 5479, July 28 2000, p. 596–590.

De Waal, F. *Peacemaking Among Primates*, Harvard University Press, 1990.

Nuzzo, R. "Profile of Frans B. M. de Waal," *PNAS*, v. 102, 32, August 9 2005, p. 11137–11139.

7: THE FACE IN THE DESERT

Gibbons, A. *The First Human*, Doubleday, 2006, p. 3–4.

Lieberman, D. *The Story of the Human Body*, Vintage, 2013, p. 32, fig 2.

Wood, B. "Paleoanthropology: Hominid Revelations from Chad," *Nature*, v. 418, July 11 2002, p. 133–135.

Zollikofer, P. E. et al. "Virtual Cranial Reconstruction...," *Nature*, v. 434, April 7, 2005, p. 755–759.

8: UNEXPECTED ARDI: A FOSSIL IN FRAGMENTS

DeSilva, J. "Where Do We Begin?," *Natural History*, v. 126, September 2018, p. 10–12.

Gibbons, A. "A New Kind of Ancestor: Ardipithecus Unveiled," *Science*, v. 326, October 2 2009, p. 36–40. See also scientific papers by Tim White and colleagues in this special issue.

Lieberman, D. *The Story of the Human Body*, Vintage, 2013.

9: LUCY, THE ICONIC ANCESTOR

Johanson, D. "The Paleoanthropology of Hadar, Ethiopia," *Comptes Rendu Palevol*, v. 16, 2, 2017, p. 140–154.

Johanson D., & Edey, M. *Lucy: The Beginnings of Humankind*, Simon & Schuster, 1981.

Kimbel, W., & Delezene, L. "Lucy Redux," *Yearbook of Physical Anthropology*, v. 52, 2009, p. 2–48.

Wynn, J. G. et al. "Diet of *Australopithecus afarensis...*," *PNAS*, v. 110, 26, June 25 2013, p. 10495–10500.

10: FIRST FOOTPRINTS

Hatala, G. et al. "Laetoli Footprints Reveal a Bipedal Gait...," *Philosophical Transactions of the Royal Society B*, August 3 2016.

Johanson, D., & Edgar, B. *From Lucy to Language*, Simon & Schuster, 2006, p. 142.

Jungers, W. J. "These Feet Were Made for Walking," *eLife Sciences*, December 14 2016.

Masao, F. T. et al. "New Footprints from Laetoli," *eLife Sciences*, December 14 2016.

11: THE PEACE CHILD

Alemseged, Z. et al. "A Juvenile Early Hominin Skeleton from Dikika, Ethiopia," *Nature*, v. 443, 7109, September 21 2006, p. 296–301.

Maxmen, A. "Digging through the World's Earliest Graveyard," *Nautilus* website, September 25 2014.

Sloan, C. P. "Childhood Origins," *National Geographic*, November 2006.

Wong, K. "Lucy's Baby," *Scientific American* website, September 20 2006.

12: A TALE OF TWO NUTCRACKERS

Keyser, A. "New Finds in South Africa," *National Geographic*, May 2000, p. 76–82.

Keyser, A. "The Drimolen Skull...," *South African Journal of Science*, v. 96, April 2000, p. 189–93.

Ungar, P. S. *Evolution's Bite*, Princeton University Press, 2017, p. 67.

Ungar, P. S. "The Real Paleo Diet," *Scientific American*, July 2018, p. 46.

13: SEARCHING FOR THE EARLIEST TOOLS

Braun, D. R. et al. "Earliest-Known Oldowan artifacts..." *PNAS*, v. 115, 24, June 11 2019, p. 11712–11717.

De Lumley, H. et al. "The First Technical Sequences...," *Antiquity*, v. 92, 365, 2018, p. 1151–1164.

Drake, N. "Wrong Turn Leads to Discovery of Oldest Stone Tools," *National Geographic* website, May 20 2015.

Harmand, S. et al. "3.3-Million-Year-Old Stone Tools...," *Nature*, v. 521, May 21 2015, p. 310–315.

Semaw, S. et al. "2.6-Million-Year-Old Stone Tools...," *Journal of Human Evolution*, v. 45, 2003, p. 169–177.

14: CHIMPANZEES OF THE TAÏ FOREST

Boesch C. et al. "Technical Intelligence and Culture: Nut Cracking in Humans and Chimpanzees," *American Journal of Physical Anthropology,* v. 163, 2017, p. 339–355.

Boesch, C., & Boesch, H. "Hunting Behavior of Wild Chimpanzees in the Taï National Park," *American Journal of Physical Anthropology,* v. 78, 1989, p. 547–573.

Boesch, C., & Boesch-Achermann, H. "Dim Forest, Bright Chimps," *Natural History*, September 1991, p. 74–79.

15: CHIMPANZEE HUNTING

Pruetz, J. D. et al. "New Evidence on the Tool-Assisted Hunting...," *Royal Society Open Conference,* v. 2, April 1 2015, p. 140507.

16: BONOBOS' POWERFUL SISTERHOOD

De Waal, F. *The Bonobo and the Atheist,* Norton, 2014.

De Waal, F. "Bonobo Sex and Society," *Scientific American,* June 1 2006.

De Waal, F., & Lanting, F. *Bonobo: The Forgotten Ape*, University of California Press, 1998.

17: FIRST OF OUR FAMILY

"Discovery at Ledi-Geraru," *Origin Stories*, The Leakey Foundation podcast, episode 5, 2015.

Foley, R. et al. "Major Transitions in Human Evolution," *Philosophical Transactions of the Royal Society B*, v. 371, 2016, p. 20150229.

Kimbel, W., & Villmoare, B. "From *Australopithecus* to *Homo*: The Transition that wasn't," *Philosophical Transactions of the Royal Society B*, v. 371, 2016, p. 20150248.

Villmoare, B. et al. "Early *Homo* at 2.8...," *Science*, v. 347, March 20 2015, p. 1352–1354.

18: THE RISE AND FALL OF 1470

Hawks, J. "The Plot to Kill *Homo habilis,*" *Medium* website, March 20 2017.

Leakey, R. E., & Lewin, R. *Origins,* E. P. Dutton, 1978.

Morell, V. *Ancestral Passions*, Simon & Schuster, 1995.

Wood, B. "Fifty Years after *Homo habilis*," *Nature*, v. 508, April 3 2014, p. 31–33.

19: THE SKULLS IN THE CELLAR

Gibbons, A. "Meet the Frail, Small-Brained People Who First Trekked Out of Africa," *Science* website, November 2 2016.

Lordkipanidze, D. "Postcranial Evidence from Early *Homo* from Dmanisi, Georgia," *Nature,* v. 449, 2007, p. 305–310.

Tarlach, G. "The First Humans to Know Winter," *Discover* magazine website, February 25 2015.

Wong, K. "Stranger in a New Land," *Scientific American* website, June 1 2006.

20: THE CASE OF THE SHRINKING ANCESTOR

Cunningham, D. L. et al. "The Effects of Ontogeny...," *Journal of Human Evolution,* v. 121, August 2018, p. 119–127.

Langdon, J. H. *The Science of Human Evolution,* Springer, 2016. See Case Study 15.

Walker, A., & Leakey, R.E. *The Nariokotome Homo Erectus Skeleton*, Harvard University Press, 1993.

21: BORN, AND EVOLVED, TO RUN

Bramble, D. M., & Lieberman, D. E. "Endurance Running and the Evolution of *Homo," Nature*, v. 432, November 18 2004, p. 345–352.

Chen, I. "Born to Run," *Discover* magazine website, May 28 2006.

Lieberman, D. E. "Born and Evolved to Run," *Origin Stories*, Leakey Foundation podcast, episode of July 19 2016.

Lieberman, D. E. *The Story of the Human Body,* Vintage, 2013, p. 85–88.

22: THE GREAT ANCESTRAL BAKE-OFF

Dibble, H. L. et al. "How Did Hominins Adapt to Ice Age Europe without Fire?," *Current Anthropology,* v. 58, 16, August 2017, p. S278.

Gorman, R. M. "Cooking Up Bigger Brains," *Scientific American* website, January 1 2008.

Gowlett, J.A.G. "The Discovery of Fire by Humans...," *Philosophical Transactions of the Royal Society B*, v. 371, 1696, January 18 2016, p. 20150164.

Wrangham, R. *Catching Fire*, Basic Books, 2009.

Wrangham, R. "Control of Fire in the Paleolithic," *Current Anthropology*, v. 58, 16, August 2017, p. S305.

23: MOUNTAIN GORILLA MYSTERY

"Stacy Rosenbaum: A Primatologist's Journey," Dian Fossey Gorilla Fund website, November 9 2011.

Rosenbaum, S. et al. "Infant Mortality Risk," *PLOS One*, February 10 2016.

Rosenbaum, S. et al. "Male Rank, not Paternity...," *Animal Behaviour,* v. 104, June 2015, p. 13–24.

24: GORILLA BONE DETECTIVES

Ault, A. "Dian Fossey's Gorilla Skulls...," *Smithsonian* website, March 17 2017.

Galbany, J. et al. "Body Growth and Life History...," *American Journal of Physical Anthropology,* v. 163, 2017, p. 570–590.

Thompson, J. L., & Nelson, A. "Middle Childhood and Modern Human Origins," *Human Nature*, v. 223, September 2011, p. 249–80.

25: SECRETS OF THE ORANGUTAN CYCLE

Cohen, J. "Orangutan Genome Is Full of Surprises," *Science* website, January 26 2011.

Knott, C. "Orangutans in the Wild," *National Geographic,* August 1998, p. 36–42.

Vogel, E. et al. "Orangutans on the Brink of Protein Bankruptcy," *Biology Letters*, v. 8, December 14 2012, p. 333–336.

Voigt, M. et al. "Global Demand for Natural Resources Eliminated More than 100,000...," *Current Biology,* v. 28, 5, March 5 2018, p. 761.

White, M. "Out on a Limb," *National Geographic*, December 2016, p. 56–78.

26: AMBUSH AT THE LAKE

Balter, M. "The Killing Ground," *Science*, v. 344, 2014, p. 1080–1083.

Conard, N. J. et al. "Excavations at Schöningen and Paradigm Shifts in Human Evolution," *Journal of Human Evolution*, v. 89, 2015, p. 3.

Milks, A. et al. "External Ballistics of Pleistocene Hand-Held Spears...," *Nature Scientific Reports*, v. 9, 2019, p. 820.

Thieme, H. "Lower Paleolithic Hunting Spears from Germany," *Nature*, v. 385, 27, 1997, p. 807–810.

27: OUR AFRICAN MOTHERS

Cann, R. L. et al. "Mitochondrial DNA and Human Evolution," *Nature*, v. 325, 1987, p. 31–36.

Harris, E. E. *Ancestors in Our Genome*, expanded paperback edition, Oxford University Press, 2019.

Langdon, J. H. *The Science of Human Evolution,* Springer, 2016.

Reich, D. *Who We Are and How We Got There*, Pantheon, 2018.

Relethford, J. *Reflections of Our Past,* Routledge, 2018.

28: THE EARLIEST "US"

"Rewriting Our Story," *Origin Stories,* The Leakey Foundation podcast, episode 16, 2008.

Hawks, J. "Three Big Insights into our African Origins," *Medium* website, January 3, 2019.

Hublin, J. J. et al. "New Fossils from Jebel Irhoud, Morocco," *Nature,* v. 546, 7657, 2017, p. 289–292.

Wong, K. "Ancient Fossils from Morocco..." *Scientific American* website, June 8 2017.

29: THE ORIGINAL SOCIAL NETWORK

Brooks, A. et al. "Long-Distance Stone Transport and Pigment Use...," *Science*, March 15 2018, p. 90–94.

Gibbons, A. "Complex behavior arose at dawn of humans," *Science*, v. 359, 6381, March 16 2018, p. 1200–1201.

Potts, R. et al. "Environmental Dynamics...," *Science*, v. 360, April 6 2018, p. 86–90.

30: THE PIT OF THE BONES

Arsuaga, J. L. et al. "Postcranial Morphology of the Middle Pleistocene Humans...," *PNAS*, v. 112, 37, September 15 2015, p. 11524–11529.

Arsuaga, J. L. et al. "Sima de los Huesos...," *Journal of Human Evolution*, v. 1197, 33, August 1997, p. 109–127.

Gibbons, A. "Fossils Put a New Face...," *Science* website, June 19 2014.

Perez, P-J. et al. "Paleopathological Evidence of the Cranial Remains...," *Journal of Human Evolution,* v. 33, 1997, p. 409–421.

31: RESOURCEFUL "BRUTES": THE PEOPLE OF THE ROCK

Finlayson, C. et al. "Birds of a Feather: Neanderthal Exploitation of Raptors and Corvids," *PLOS ONE,* 7, 9, September 17 2012, p. 1–9.

Finlayson, C. et al. "Late Survival of Neanderthals...," *Nature*, v. 443, 7113, October 19 2006, p. 850–853.

Finlayson, C. *The Smart Neanderthal*, Oxford University Press, 2019.

Hofmann, D. L. et al. "U-Th Dating of Carbonate Crusts..." *Science*, v 59, 6378, February 23 2018, p. 912–915.

Rodriguez-Vidal, J. et al. "A Rock Engraving Made by Neanderthals in Gibraltar," *PNAS*, v. 111, 37, September 16 2014, p. 13301–13306.

Wong, K. "Secrets of Neanderthal Cognition Revealed," *Scientific American* website, February 1 2015.

32: CRACKING THE NEANDERTHAL CODE

Gibbons, A. "Neanderthals and Moderns Make Imperfect Mates," *Science*, v. 343, 6177, 2014, p. 471–472.

Green, R. E. et al. "A Draft Sequence of the Neanderthal Genome," *Science*, v. 328, May 7 2010, p. 710–722.

Pååbo, S. *Neanderthal Man*, Basic Books, 2014.

Reich, D. *Who We Are and How We Got Here*, Pantheon, 2018.

Relethford, J. *Reflections of Our Past,* Routledge, 2018.

33: THE FINGER IN THE CAVE

Gibbons, A. "Our Mysterious Cousins...," *Science* website, March 29 2019.

Krause, J. et al. "The Complete Mitochondrial DNA Genome...," *Nature*, v. 464, 7290, 2010, p. 894–897.

Reich, D. *Who We Are and How We Got Here*, Pantheon, 2018.

Shreeve, J. "Case of the Missing Ancestor," *National Geographic*, July 2013.

34: THE FATE OF THE HYBRIDS

Fu, Q. et al. "An Early Modern Human from Romania...," *Nature*, v. 524, August 13 2015, p. 216–219.

Gibbons, A. "Our Mysterious Cousins...." *Science* website, March 29 2019.

Mellars, P., & French, J. C. "Ten-Fold Population Increase...," *Science*, v. 333, July 29 2011, p. 623–637.

Rios, L. et al. "Skeletal Anomalies in the Neandertal Family...," *Scientific Reports*, v. 9, 1697, February 8 2019.

Slon, V. et al. "The Genome of the Offspring of a Neanderthal...," *Nature*, v. 561, August 22 2018, p. 113–116.

Trinkaus, E. et al. "Early Modern Cranial Remains...," *Journal of Human Evolution*, v. 45, 2003, p. 245–253.

Warren, M. "Mum's a Neanderthal...," *Nature* website, August 22 2018.

35: ENIGMA OF THE LION MAN

Conard, N. "A Female Figurine from the Basal Aurignacian...," *Nature*, v. 459, 2009, p. 248–252.

Conard, N. J. et al. "New Flutes Document...," *Nature*, v. 460, June 24 2009, p. 737–740.

Dutkiewicz, E., & Conard, N. "The Symbolic Language of the Swabian Aurignacian," *PALEO*, 2016, p. 149–164.

Kind, C.-J. et al. "The Smile of the Lion Man," *Quartar*, v. 61, 2014, p. 129–145.

36: THE BEASTS OF CHAUVET

Chauvet, J-M. et al. *Chauvet Cave: The Discovery of the World's Oldest Paintings,* Thames & Hudson, 1996.

Clottes, J., & Azema, M. "Felines of the Chauvet Cave," *Bulletin de la Société Préhistorique Française*, v. 102, 1, January 2005, p. 173–182.

Pettitt, P., & Bahn, P. "An Alternative Chronology...," *Antiquity,* v. 89, 345, June 2015, p. 542–553.

Quiles, A. et al. "A High-Precision Chronological Model...," *PNAS*, v. 113, 17, April 26 2016, p. 4670–4675.

37: GRANDMOTHERS AND OTHERS

"The Grandmother Hypothesis," *Origin Stories,* The Leakey Foundation podcast, episode 15.

Coxworth, J. E. et al. "Grandmothering Life Histories...," *PNAS*, v. 112, 38, September 22 2015, p. 11806–11811.

Hawkes, K., & Coxworth, J. E. "Grandmothers and the Evolution of Human Longevity," *Evolutionary Anthropology*, v. 226, 2013, p. 294–302.

Hawkes, K., O'Connell, J. F. et al. "Grandmothering, Menopause, and the Evolution of Human Life Histories," *PNAS*, v. 95, August 1998, p. 1336–1339.

Hrdy, S. *Mothers and Others: The Evolutionary Origins of Mutual Understanding*, Harvard University Belknap Press, 2011.

Hrdy, S. *Mother Nature*, Pantheon, 1999.

38: BABOONS' PERILOUS CHILDHOOD

Alberts, S. C. "The Challenge of Survival for Wild Infant Baboons," *American Scientist*, v. 104, 6, 2016, p. 366–373.

Pennisi, E. "Baboon Watch," *Science*, v. 346, October 17 2014, p. 293–295.

Tung, J. et al. "Cumulative Early Life Adversity...," *Nature Communications,* v. 7, 11181, April 19 2016.

39: INSIDE BABOON MINDS

"Being a Nice Animal," *Origin Stories*, The Leakey Foundation podcast, episode 06.

Cheney, D.L. & Seyfarth, R.M. *Baboon Metaphysics*, University of Chicago Press, 2007.

Cheney, D.L. & Seyfarth, R.M. "Cognition, Communication, and Language," *Evolutionary Anthropology*, v. 21, 2012, p. 186.

40: THE HOBBIT ENIGMA

NOVA, *Alien From Earth*, PBS website, November 11 2008.

Brown, P. et al. "A New Small-Bodied Hominin...," *Nature,* v. 431, 1029, October 2004, p. 1055–1061.

Dalton, R. "Little lady of Flores...," *Nature,* v. 431, October 28 2004, p. 1029.

Detroit, F. et al. "A New Species of *Homo...," Nature*, v. 568, April 10 2019, p. 181–186.

Falk, D. *The Fossil Chronicles,* University of California Press, 2011.

Morwood, M. *A New Human*, Harper Collins, 2007.

Wong, K. "Rethinking the Hobbits," *Scientific American*, v. 22, 1s, December 2012, special editions, p. 84–91.

41: THE CHAMBER OF STARS MYSTERY

NOVA/National Geographic, *Dawn of Humanity*, PBS website, September 16 2015.

"Rising Star," *Origins Stories,* The Leakey Foundation podcast, episode 26.

Berger, L. R. et al. "*Homo naledi,* a New Species...," *eLife*, September 10 2015.

Dirks, P. et al. "The Age of *Homo naledi*...," *eLife*, May 9 2017.

42: THE ROOTS OF WAR

Ferguson, R. B. "War is *Not* Part of Human Nature...," *Scientific American,* v. 319, 3, September 2018, p. 76–81.

Gat, A. "Proving Communal Warfare among Hunter-Gatherers...," *Evolutionary Anthropology,* v. 24, 2015, p. 111–126.

Gomez, J. M. et al. "The Phylogenetic Roots of Human Violence," *Nature,* v. 358, October 13 2016, p. 233–238.

Lahr, M. "Mirazón Lahr *et al* Reply," *Nature,* v. *539,* November 23 2016, E10–E11.

Lahr, M. et al. "Inter-Group Violence...," *Nature*, v. 529, January 21 2016, p. 394–398.

Pinker, S. *The Better Angels of Our Nature*, Viking, 2011.

Sala, N. et al. "The Sima de los Huesos Crania..." *Journal of Archaeological Science,* v. 72, August 2016, p. 25–43.

Wrangham, R. *Demonic Males,* Houghton Mifflin Harcourt, 1996.

Zollikofer, C. P. E. et al. "Evidence for Interpersonal Violence in the St. Césaire Neanderthal," *PNAS*, v. 99, 9, April 30 2002, p. 6444–6448.

43: SKIN DEEP: THE ENIGMA OF SKIN COLOR

Gibbons, A. "Shedding Light on Skin Color," *Science*, v. 346, November 21 2014, p. 934–936.

Jablonski, N. G. *Skin, A Natural History,* University of California Press, 2013.

Lalueza-Fox, C. et al. "A Melanocortin 1 Receptor Allele...," *Science*, v. 318, 5855, November 30 2007, p. 1453–1455.

Quillen, E.E. et al. "Shades of Complexity...," *American Journal of Physical Anthropology*, 2018, p. 1–23.

44: LETHAL HEIGHTS: SURVIVING THIN AIR

NOVA, *Secrets of the Sky Tombs*, PBS website, October 31 2018.

Beall, C. "Adaptation to High Altitude," *Annual Review of Anthropology*, v. 43, 2014, p. 251–272.

Bigham, A., & Lee, F. "Human High Altitude Adaptation...," *Genes and Development*, v. 28, September 20 2015, p. 2189–2204.

Chen, F. et al. "A Late Middle Pleistocene Denisovan Mandible...," *Nature*, v. 569, 7756, May 1 2019, p. 409–412.

Jeong, C. et al. "Long-Term Genetic Stability...," *PNAS*, v. 113, 27, July 5 2016, p. 7485–7490.

Lorenzo, F. R. et al. "A Genetic Mechanism for Tibetan...," *Nature Genetics,* v. 49, August 17 2014, p. 951–956.

45: DEADLY COMPANIONS

Bos, K. I. et al. "Pre-Columbian Mycobacterial Genomes Reveal Seals...," *Nature*, v. 514, 7523, October 24 2014, p. 494–497.

Brandt, G. et al. "Settlement Burials at the Karsdorf LBK site," *Proceedings of the British Academy*, v. 198, 2014, p. 95–114.

Lieberman, D. *The Story of the Human Body*, Vintage, 2013.

Reich, D. *Who We Are and How We Got Here,* Pantheon, 2018.

Shriner, D., & Rotimi, N. "Whole-Genome-Sequence-Based Haplotypes...," *American Journal of Human Genetics*, v. 102, April 5 2018, p. 547–556.

46: THE AFRICAN MILK TRAIL

Dunn, R. "Follow the Drinking Gourd...," *Natural History*, v. 119, May 1 2011, p. 5.

Ranciaro, A. et al. "Genetic Origins of Lactase Persistence...," *American Journal of Human Genetics,* v. 94, 4, April 3 2014, p. 496–510.

Segurel, L., & Bon, C. "On the Evolution of Lactase Persistence...," *Annual Review of Genomics and Human Genetics,* v. 18, 2017, p. 297–319.

Tishkoff, S. A. et al. "Convergent Adaptation...," *Nature Genetics*, v. 39, 1, December 10 2007, p. 31–40.

47: THE STRANGE RITUALS OF CAPUCHINS

Perry, S. *Manipulative Monkeys: The Capuchins of Lomas Barbudal*, Harvard University Press, 2011.

Perry, S. "Social Traditions and Social Learning in Capuchin Monkeys," *Philosophical Transactions of the Royal Society B*, v. 366, 1567, April 12 2011, p. 988–996.

48: KILLING THE INFANTS

Borries, C. et al. "DNA Analyses Support...," *Proceedings of the Royal Society* B, v. 266, 1422, May 7 1999, p. 901–904.

Hrdy, S. *Mothers and Others: The Evolutionary Origins of Mutual Understanding*, Harvard University Belknap Press, 2011.

Hrdy, S. *Mother Nature*, Pantheon, 1999.

Hrdy, S. H. "Infanticide as a Primate Reproductive Strategy," *American Scientist*, v. 65, 1, January–February 1977, p. 40–49.

Packer, C. "Infanticide is No Fallacy," *American Anthropologist,* v. 102, 4, December 2000, p. 829–831.

49: MONKEY MONOGAMY

Edgar, B. "Our Secret Evolutionary Weapon," *Scientific American* website, September 1 2014.

Fernandez-Duque, E. "Love in the Time of Monkeys," *Natural History,* December 2014/January 2015, p. 16–21.

Fernandez-Duque, E. "Of Monkeys, Moonlight, and Monogamy...," in Strier, K., ed., *Primate Ethnographies,* Taylor and Francis, 2016.

Fernandez-Duque, E. "Social Monogamy in Wild Owl Monkeys...," *American Journal of Primatology,* v. 78, 3, March 2015, p. 355–371.

Fernandez-Duque, E. & Huck, M. "Till Death or an Intruder...," *PLOS ONE*, v. 8, 1, January 2013, e53724.

Heinrich, J. et al. "The Puzzle of Monogamous Marriage," *Philosophical Transactions of the Royal Society*, v. 367, 1598, March 5 2012, p. 657–669.

Huck, M. et al. "Correlates of Genetic Monogamy...," *Proceedings of the Royal Society B*, May 7 2014, p. 1–8.

50: MAPPING THE GENOME: OUR UNFINISHED JOURNEY

General discussions of our future evolution

Harris, E. E. *Ancestors in Our Genome*, Oxford University Press, 2019, chapt. 7.

Hawks, J. "Humans Never Stopped Evolving," *The Scientist*, August 1 2016.

Hawks, J. "No, Humans Have Not Stopped Evolving," *Scientific American,* September 2014.

Lee, S-H. Close Encounters with Humankind, W. W. Norton, 2017, p. 236.

Leiberman, D. *The Story of the Human Body*, Vintage, 2013, chapt. 7.

Pritchard, J. K. "How We Are Evolving," *Scientific American* website, October 1 2010.

Reich, D. *Who We Are and How We Got Here*, Pantheon, 2018, chapt. 12.

Rutherford, A. *A Brief History of Everyone Who Ever Lived,* 2016, chapt. 8.

Stringer, C. *Lone Survivors*, Times Books, 2011, p. 268–278.

Ward, P. "What May Become of Homo sapiens," *Scientific American*, December 2012.

Other sources for "Mapping the Genome"

Estrada, A. et al. "Impending Extinction Crisis...," *Science Advances*, v. 3, 1, January 18 2017, e1600946.

Fu, W. et al. "Analysis of 6,515 Exomes...," *Nature*, v. 493, January 10 2013, p. 216–220.

Hawks, J. et al. "Recent Acceleration...," *PNAS,* v. 104, 2, December 26 2007, p. 20753–20758.

Kühl, H. S. et al. "Human impacts erode...," *Science*, v. 363, 6434, March 29 2019, p. 1453–1455.

Voigt, M. et al. "Global Demand for Natural Resources...," *Current Biology*, v. 28, 5, March 5 2018, p. 761–769.

Zuk, M. *Paleofantasy*, W. W. Norton, 2013, chapt. 5.

PHOTOGRAPH CREDITS

page number/credit

DUST JACKET

front Copyright, reconstruction, Elisabeth Daynes; photo, Sylvain Entressangle

flap Courtesy of the author

FOREWORD

vii Courtesy Kirk Johnson, photo by Piers Leigh

PREFACE

x Courtesy The Leakey Foundation Archive, 1975

xi Courtesy Rachael Rothstein, 2018

xii Drew Altizer Photography

INTRODUCTION

1 Robert Sisson/Nat Geo Image Collection

2 *Left:* Bettmann/Getty Images
Right: Science History Images/Alamy Stock Photo

3 Rudoph F. Zallinger

4 *Top:* The Simpsons' copyright and trademark 1990 Twentieth Century Fox Film Corporation. All rights reserved
Bottom: Public Domain

5 Public Domain

6–7 Dr. Prateek Lala for The Leakey Foundation

8 Max Planck Institute for Evolutionary Anthropology

9 Science History Images/Alamy Stock Photo

10 *Left:* courtesy The Leakey Foundation Archive
Right: courtesy The Leakey Foundation Archive

11 Robert Sisson/Nat Geo Image Collection

12 Courtesy The Leakey Foundation Archive

13 *Top:* Hugo van Lawick/Nat Geo Image Collection
Bottom left: Robert I. M. Campbell/Nat Geo Image Collection
Bottom right: courtesy Rodney Brindamour/Orangutan Foundation International

14 JGI/Hugo van Lawick

15 *Left:* courtesy The Leakey Foundation Archive
Right: copyright Jane Goodall Institute/by Chase Pickering

16 JGI/Hugo van Lawick

17 Copyright Kathelijne Koops

18 Courtesy Bob Campbell Papers, African Studies Collections, Special and Area Studies Collections, George Smathers Libraries, University of Florida, Gainesville, Florida

19 *Top:* courtesy Bob Campbell Papers, African Studies Collections, Special and Area Studies Collections, George Smathers Libraries, University of Florida, Gainesville, Florida
Bottom: courtesy Bob Campbell Papers, African Studies Collections, Special and Area Studies Collections, George Smathers Libraries, University of Florida, Gainesville, Florida

20 Album/Alamy Stock Photo

21 Courtesy Bob Campbell Papers, African Studies Collections, Special and Area Studies Collections, George Smathers Libraries, University of Florida, Gainesville, Florida

22 Copyright Jeff Mauritzen

24 Courtesy The Leakey Foundation Archive

25 MARKA/Alamy Stock Photo

26 ANL/Shutterstock

27 Gordon Gahan/Nat Geo Image Collection

28 Copyright Camerapix/Science Source

29 Copyright John Reader/Science Source

30 Copyright Des Bartlett/Science Source

32 Marion Kaplan/Alamy Stock Photo

33 Copyright Marion Kaplan

34 Copyright Kenneth Garrett

37 Courtesy Patick Nduru Gathogo

38 Copyright John Reader/Science Source

39 Robert Sisson/Nat Geo Image Collection

SUPPORTING A NEW GENERATION OF SCIENTISTS

41 Courtesy Liang Bua Team

42 *Left:* courtesy Felix Kisena @ Penina Emanuel
Right: courtesy Felix Kisena@ Penina Emanuel

43 *Top:* courtesy Dereje Bewket
Bottom: courtesy Sharmi Sen

44 Courtesy Onja Razafindratsima

45 Courtesy Hesham Sallam

46 Courtesy David Fuller

DISCOVERING US

49 Dr. Xijun Ni

50 Natalia–Stock.adobe.com

51 Copyright J. H. Matternes

52 Isaiah Nengo photo Christopher Kiare

54 Paul Tafforeau (ESRF)

55 Isaiah Nengo

56 David Haring, Duke Lemur Center

58 Copyright Peter Schouten

61 Roger Ressmeyer/Corbis/VCG

62 John Mitani

65 Konrad Wothe/NPL

67 Copyright Ronan Donovan

68 Copyright Ronan Donovan

70 Alain Beauvillain, University of Paris x Nanterre

72 Patrick Robert Corbis/via Getty Images

73 Alain Beauvillain, University of Paris x Nanterre

75 Copyright J. H. Matternes

77 Copyright 2009, Tim D. White, humanoriginsphotos.com

78 Copyright 1994, David L. Brill, humanoriginsphotos.com

80 Copyright 2021, John Gurche

81 AKG-images

82–83 Robert I. M. Campbell/Nat Geo Image Collection

84 John Reader/Science Source

87 Copyright Reconstruction Elisabeth Daynes, photo E. Daynes

88 Copyright Kenneth Garrett

90 Copyright Kenneth Garrett

93 Photo Kathy Schick and Nicholas Toth, courtesy of the Stone Age Institute

94–95 Dr. Michael J. Rogers, Southern Connecticut State University

96–97 Dr. Michael J. Rogers, Southern Connecticut State University

101 Liran Samuni, Tai Chimpanzee Project

102 Frans Lanting/Nat Geo Image Collection

105 McKensey Miller, Fongoli Savanna Chimpanzee Project

106 Takeski Furuichi

110 John Rowan, University of Albany

113 William Kimbel, Institute of Human Origins, Arizona State University

114 John Reader/Science Photo Library

118 Copyright, reconstruction Elisabeth Daynes;
photo, Sylvain Entressangle

120 Computer reconstruction of the five Dmanisi skulls (Marcia Ponce de León and Christoph Zollikofer, University of Zurich)

121 Kenneth Garrett/Nat Geo Image Collection

123 Copyright Kennis & Kennis

124 Copyright 1985, David L. Brill

126 David Ramos Oviedo

129 Copyright Izla Photography, Izla Bethdavid Boltena

130 Photostock–Israel/Alamy Stock Photo

132 Eyal Bartow/Alamy Stock Photo

135 W. Eckardt/Dian Fossey Gorilla Fund

136 J. Burbridge/Dian Fossey Gorilla Fund

139 Arco Images Gmblt/Alamy Stock Photo

140 Courtesy of the Mountain Gorilla Skeletal Project

143 Copyright Tim Laman

144 Copyright Tim Laman

146 Copyright Peter Pfarr/Lower Saxony State Office of Cultural Heritage
149 Copyright Kenneth Garrett
151 C. Rottensteiner, Wikimedia Commons CC BY SA 3.0
153 Martin Harvey/Alamy Stock Photo
154 Shannon McPherron, MPI-EVA Leipzig CC BY SA 2.0
155 Mark Steinmetz/Visum
156 Copyright Phillip Gunz MPI EVA Leipzig CC BY SA 2.0
157 Copyright Kennis & Kennis
158 Copyright Jason F. Nichols
163 Javier Trueba/MSF/Science Photo Library
164 Javier Trueba/MSF/Science Photo Library
165 Javier Trueba/MSF/Science Photo Library
167 Javier Trueba/MSF/Science Photo Library
169 *Left:* Stewart Finlayson
Right: Luka Mjeda
170 Frank Vinken for Max Planck Society
173 Frank Vinken for Max Planck Society
175 Robert Clark/Nat Geo Image Collection
176 Robert Clark/Nat Geo Image Collection
179 Copyright MPI f. Evolutionary Anthropology
181 Erik Trinkaus and João Zihão
182 Photo by Yvonne Muhles copyright State Office of Cultural Heritage Baden-Wuerttemberg/Museum Ulm
184 Photo by Hilde Jensen, copyright University of Tuebingen
185 AP Photo/Daniel Maurer
187 Philippe Morel
188–89 Steven Alvarez/National Geographic Creative
191 Brian Wood
192 Fernando A. Campos
194 Courtesy Susan Alberts photo by Joshua See
196–97 Jurgen and Christine Sohns/Alamy Stock Photo
198 Keena Roberts
200–01 Liang Bua Team
203 Yousuke Kaifu
204 Robert Clark/Nat Geo Image Collection
207 Robert Clark/Nat Geo Image Collection
209 Robert Foley
210 Marta Mirazón Lahr
213 Copyright Angelica Dass/Humanae
216 Copyright Cat Vinton
218 Courtesy Cory Richards/National Geographic
219 Dongju Zhang/Lanzhou University Wikipedia, CC BY SA 4.0
221 CDC/Dr. Erskine Palmer
222 *Top:* CDC/NIAID
Bottom left: Dr. Gopal Murti/Science Photo Library
Bottom right: State Office of Heritage Management and Archaeology Saxony-Anhalt, Nicole Nicklisch
224 Copyright Caroline Irby
227 Dr Alessia Ranciaro
229 Susan Perry
230 Copyright Jim Fenton
233 Copyright Biosphoto/Cyril Ruoso
234 Courtesy Sarah Hrdy, photo by D. B. Hrdy/Anthro-Photo
236 Emilio White/Owl Monkey Project, Formosa Argentina
239 Margaret Corley/Owl Monkey Project, Formosa Argentina
240 aAron munson, Handful of Films Inc.
243 Paul Nicklen/Nat Geo Image Collection
244 Dr. Gwen Duytschaever
247 Pradiptajati Kusuma
248 Kasiisi Project
249 Zarin Machandra
251 Melville B. Grosvenor/Nat Geo Image Collection

INDEX

ABOUT THE AUTHOR

Evan Hadingham is a science journalist and the author of books about archaeology, prehistoric art, human evolution, and the history of aviation. His love of archaeology began as a teenager and he later volunteered on archaeological rescue digs in Britain and France. For his book, *Secrets of the Ice Age*, he researched and photographed prehistoric cave art throughout France and Spain. After a graduate degree in prehistory and archaeology at the University of Sheffield, Hadingham moved to his current home in Massachusetts. He has helped carry out survey work on ancient Maya sites and investigated the famous Nazca lines in his book, *Lines to the Mountain Gods.* For three decades, he has worked in broadcast journalism at WGBH Educational Foundation in Boston, producers of the PBS science series NOVA, where he is currently Senior Science Editor. He has helped oversee the production of many award-winning NOVA documentaries and his feature articles have appeared in *National Geographic*, *Smithsonian*, *The Atlantic*, and *Discover*.